William Besant

Elementary hydrostatics

With chapters on the motion of fluids and on sound

William Besant

Elementary hydrostatics
With chapters on the motion of fluids and on sound

ISBN/EAN: 9783337276683

Printed in Europe, USA, Canada, Australia, Japan

Cover: Foto ©berggeist007 / pixelio.de

More available books at **www.hansebooks.com**

ELEMENTARY

HYDROSTATICS

WITH CHAPTERS ON THE MOTION OF FLUIDS
AND ON SOUND.

BY

W. H. BESANT, Sc.D., F.R.S.

FELLOW OF ST JOHN'S COLLEGE, CAMBRIDGE.

SEVENTEENTH EDITION.

Ἄριστον μὲν ὕδωρ.

LONDON
GEORGE BELL AND SONS
1895

PREFACE.

THE object of this Treatise is to place before the student a complete series of those propositions in Hydrostatics, the solutions of which can be effected without the aid of the Differential Calculus, and to illustrate the theory by the description of many Hydrostatic Instruments, and by the insertion of a large number of examples and problems.

In doing this I have had in view the courses of preparation necessary for the first three days of the Examination for the Mathematical Tripos, for some of the Examinations of the University of London, of the Science and Art Department of the Committee of Council on Education, of the Civil Service Commission, and for various other Examinations in which more or less knowledge of Hydrostatics is required.

As far as possible the whole of the propositions are strictly deduced from the definitions and axioms of the subject, but it is occasionally necessary to assume empirical results, and these assumptions are distinctly pointed out.

I have thought it advisable, with a view to some of the examinations above alluded to, to give an account of some cases of fluid motion, and also to give an explanation of some of the more important phenomena of sound; in each case I have assumed, as the basis of reasoning, certain facts which can be deduced from theory by an analytical investigation, but which it may be useful to the student to accept as experimental results.

The Geometrical facts which are enunciated at the end of the Introduction are such as can be demonstrated without the aid of the Differential Calculus.

The slight historical notices appended to some of the chapters are intended to mark the principal steps in the progress of the science, and to assign to their respective authors the exact value of the advances made at different times.

For the present edition the text has been very carefully revised, and many alterations and additions have been made.

In particular the chapters on the Motion of Fluids and on Sound have been completely separated from the chapters on the equilibrium of fluids, and, in the case of each set of chapters, an uniform system of units has been maintained throughout.

I am very much indebted to Mr A. W. Flux, Fellow of St John's College, for valuable assistance in the revision of proof-sheets, and for many useful suggestions.

I may add that a new edition of the book of solutions of the examples and problems is in course of preparation, and will, I hope, shortly be published.

W. H. BESANT.

St John's College,
March, 1892.

TABLE OF CONTENTS.

viii

CHAPTER IX.

CHAPTER X.

CHAPTER XI.

CHAPTER XII.

CHAPTER XIII.

CHAPTER XIV.

CHAPTER XV.

CHAPTER XVI.

ELEMENTARY HYDROSTATICS.

INTRODUCTION.

THE object of the science of Hydrostatics is to discuss the mechanical properties of fluids, or to determine the nature of the action which fluids exert upon each other and upon bodies with which they are in contact, and to explain and classify, under general laws, the varied phenomena relating to fluids which are offered to the attention of an observer. To effect this purpose it is necessary to construct a consistent theory, founded upon observation and experiment, from which, by processes of deductive reasoning, and the aid of Geometry and Algebra, the explanations of phenomena shall flow as consequences of the definitions and fundamental properties assumed; the test of the theory will be the coincidence with observed facts of the results of such reasoning.

We shall assume in the following pages that the student is acquainted with the elements of Plane Geometry, with the simpler portions of Algebra and Trigonometry, and of Statics, and, in the later chapters, with a few of the properties of Conic Sections, and certain results of Dynamics.

In dealing with any mechanical science, we may take as the basis of our reasoning certain known laws, derived from experiment, or we may deduce these laws from a set of axioms and definitions, the axioms being the result of inductive reasonings from observed facts. With our present subject it is generally necessary to rest upon empirical laws, but in some cases these laws can be deduced from the

axiomatic definition of a fluid. For instance, in the first
chapter we have stated as experimental laws the principles
of the equality of pressure in all directions and the transmission of pressure, but this formal statement of fact is
followed by a deduction of the laws, by strict reasoning, from
the axiomatic definition.

The idea of a varying fluid pressure and of the measure
of such pressure is one of the first which presents itself as
a difficulty; the student will perceive that it is a difficulty
of the same kind as the idea of varying velocity and its
measure. A body in motion with a changing velocity has,
at any instant, a rate of motion which can be appreciated
and measured; and, in a similar manner, the pressure at any
point of a fluid can be conceived, and, by reference to proper
units, can be made the subject of calculation.

Some of the most important results of the science will
be found in the construction of Hydrostatic instruments; a
consideration of these instruments, many of which we shall
describe, will shew how universal are the practical applications of fluids, and that, while doing the hardest work of
levers and pullies, they at the same time assist in the most
delicate manipulations for determining weights and measures.
The Hydraulic Press and the Stereometer illustrate these
extreme applications of the properties of fluids.

The articles printed in smaller characters in the following chapters may if necessary be omitted during a first
reading of the subject, and the Examination papers which
follow the first eight chapters are intended as a first course
of questions upon the chapters. The examples which follow
the Examination papers are somewhat more difficult, and
should be dealt with after the former have been studied and
discussed.

The following geometrical facts are assumed and employed in some of the examples.

*The volume of a pyramid or of a cone is one-third of the
prism or cylinder on the same base and of the same altitude.*

*The volume of a sphere is $\frac{4}{3}\pi r^3$, and its surface is $4\pi r^2$,
r being the radius.*

*The volume of a paraboloid of revolution is one-half the
cylinder on the same base and of the same altitude.*

The surface of a cone is $\pi r^2 \operatorname{cosec} \alpha$, r being the radius of the base and α the semivertical angle. This may also be written $\pi r \sqrt{r^2 + h^2}$, h being the altitude of the cone.

The area of an ellipse is πab, 2a and 2b being the lengths of its axes.

The area of the portion of a parabola cut off by any ordinate is two-thirds of the rectangle, the sides of which are the ordinate and corresponding abscissa.

The area of the portion of a parabola cut off by any chord is two-thirds of the parallelogram formed by the chord, the tangent parallel to the chord, and the diameters of the parabola passing through the ends of the chord. The centroid of this area lies on the diameter through the point of contact of the tangent parallel to the chord, and divides the distance between the point of contact and the middle point of the chord in the ratio of three to two.

The area of the surface of a sphere contained between two parallel planes which intersect or touch the sphere is equal to the area, between the same planes, of the surface of the circumscribing cylinder, the axis of which is perpendicular to the planes. Also the centroid of the surface is equidistant from the planes.

The distance of the centroid of the volume of a hemisphere from the centre of the sphere is three-eighths of the radius.

CHAPTER I.

1. IT is a matter of ordinary observation that fluids are capable of exerting pressure.

A certain amount of effort is necessary in order to immerse the hand in water, and the effort is much more sensible when a light substance, such as a piece of wood or cork, is held under water, the resistance offered to the immersion being greater as the piece immersed is larger. This resistance can only be caused by the fluid pressure acting upon the surface of the body immersed.

If an aperture be made in the side of a vessel containing water, and be covered by a plate so as to prevent the escape of the water, a definite amount of force must be exerted in order to maintain the plate in its position, and this force is opposed to, and is a direct measure of, the pressure of the water.

That the atmosphere when at rest exerts pressure is shewn directly by means of an air-pump. Amongst many experiments a simple one is to exhaust the air within a receiver made of very thin glass; when the exhaustion has reached a certain point depending on the strength of the glass, the receiver will be shivered by the pressure of the external air. The action of wind, the motion of a windmill, the propulsion of a boat by means of sails, and other familiar facts offer themselves naturally as instances of the pressure of the air when in motion.

2. All such substances as water, oil, mercury, steam, air, or any kind of gas are called fluids, but in order to obtain a definition of a fluid, we have to find a property which is common to all these different kinds of substances, and which does not depend upon any of the characteristics by which they are distinguished from each other. This property is found in the extreme mobility of their particles and in the ease with which these particles can be separated from the mass of fluid and from each other, no sensible resistance being offered to the separation from a mass of fluid of a portion whether large or small.

If a very thin plate be immersed in water, the resistance to its immersion in the direction of its plane is so small as to lead to the idea that a perfectly fluid mass is incapable of exerting any tangential action, or, in other words, any action of the nature of friction, such for instance as would be exerted if the plate were pushed between two flat boards held close to each other. Observations of such experiments have led to the following definition :

A Fluid is a substance, such that a mass of it can be very easily divided in any direction, and of which portions, however small, can be very easily separated from the whole mass ;

And also to the statement of the fundamental property of a fluid, viz. :

The Pressure of a fluid on any surface with which it is in contact is perpendicular to the surface.

3. Fluids are of two kinds, liquid and gaseous, the former being practically incompressible, while the latter, by the application of ordinary force, can be easily compressed, and, if the compressing force be removed or diminished, will expand in volume.

Liquids are however really compressible, but to a slight degree.

Experiments made by Canton in 1761, Perkins in 1819,

Oersted in 1823, Colladon and Sturm in 1829, and others, have proved the compressibility of liquids.

The last two obtained the following results, employing a pressure of one atmosphere, that is 14½ lbs. on a square inch, at the temperature 0°.

Compression of unit of volume.

Mercury	·000005
Distilled water	·000049
„ „ deprived of air	·000051
Sulphuric ether	·000133

Moreover *the decrease in volume, for the same liquid, is proportional to the pressure.*

If V be the original volume of a liquid, and V' its volume under a pressure p, $V - V'$ is the decrease in the volume V, and therefore $\dfrac{V - V'}{V}$ is the decrease in each unit of volume.

Hence the law may be thus stated:

$$\frac{V - V'}{V} \propto p = \mu p,$$

where μ is different for different fluids.

Thus for mercury, if p be measured by taking one atmospheric pressure as the unit, we have $\mu = ·000005$. We shall however, in all questions relating to equilibrium, consider liquids as incompressible fluids.

Measure of fluid pressure.

4.   The pressure of a fluid on a plane is measured, when uniform over the plane, by the force exerted on an unit of area.

Thus, if a vessel with a moveable base contain water, and if it be necessary to employ a force of 60 lbs. upwards to keep the base at rest, then 60 lbs. is the pressure of the water on the base; and, supposing the area of the base to be 4 square inches, and that a square inch is the unit of area, the measure of the pressure *at* any point of the base is 15 lbs.

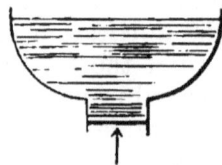

The pressure *on* a point of the base is of course zero ; the pressure *at* a point is used conventionally to express the pressure on a square unit containing the point.

If the pressure be variable over the plane, as, for instance, on the vertical side of a vessel, the pressure at any point is measured by the pressure which would be exerted on an unit of area, supposing the pressure over the whole unit to be exerted at the same rate as it is at the point.

In order to measure the pressure of a fluid at any point *within its mass*, imagine a small rigid plane placed so as to contain the point, and conceive the fluid removed from one side of the plane and the plane kept at rest by a force of P lbs. Then if α be the area of the plane, and the pressure over it be uniform, $\dfrac{P}{\alpha}$ is the pressure on each unit of area, and this is usually represented by p.

If the pressure over the plane be variable, we may suppose the area α made so small that the pressure shall be sensibly uniform, and in this case P will be small as well as α, but $\dfrac{P}{\alpha}$ or p will measure the rate of pressure at the point.

Or we may say that p, the pressure at the point, i.e. the rate of pressure per unit area, is the limiting value of $\dfrac{P}{\alpha}$, when α, and therefore P, are indefinitely diminished.

5. *Transmission of fluid pressure.*

Any pressure, applied to the surface of a fluid, is transmitted equally to all parts of the fluid.

If a closed vessel be filled with water, and if A and B be two equal openings in the top of the vessel, closed by pistons, it is found that any pressure applied at A must be counteracted by an equal pressure at B to prevent its being forced out, and if C be a piston of different size, it is found that the pressure applied at C must bear to the pressure on A the ratio of

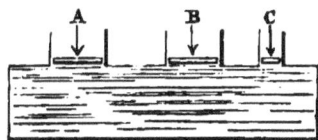

the area of C to that of A, and that this is the case whether the piston B exists or not.

Taking a more general case, if a vessel of any shape have several openings closed by pistons, kept at rest by suitable forces, it will be found that any additional force P applied to one piston will require the application, to all the other pistons, of additional forces, which have the same ratio to P as the areas of the respective pistons have to that of the piston to which P is applied.

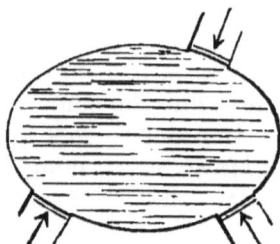

6. To explain the reason of this equal transmission, imagine a tube of uniform bore filled with water and closed by pistons at A and B. Then it may be assumed as self-evident, that any additional force applied at A will require an equal additional force at B to counteract it and keep the fluid at rest.

Now suppose in the figure that A and B are equal pistons, and draw a tube of uniform bore and of any form connecting the two, and imagine all the fluid except that contained in the tube to be solidified. This will not affect the equilibrium, inasmuch as the fluid pressure on the surface of the tube is at all points perpendicular to the surface whether the fluid be or be not solidified, and the additional pressures on A and B are equal as before.

Also, one piston (A) remaining fixed, the other (B) may be placed with its plane in any direction, and it follows that the pressure upon it is the same for all positions of its plane, or, in other words, the pressure of the fluid is the same in every direction. This proposition we shall enunciate in a general manner in the next article.

The experimental fact that the pressures on pistons of different areas are proportional to those areas may be deduced as follows.

Suppose in a closed vessel two apertures be made in which pistons are fitted, one being a square A, and the other a plane area B, formed by placing together two, three, or any number of squares equal to A; then the additional pressure on each square being equal to the additional pressure on A, the whole additional pressure will be to the additional pressure on A as the area of B is to that of A *.

7. *The pressure at any point of a fluid is the same in every direction.*

It is intended by this statement to assert that if at any point of a fluid a small plane area be placed containing the point, the pressure of the fluid upon the plane at that point will be independent of the position of the plane.

The second figure of Art. (5) will serve to illustrate the meaning of the proposition. The aperture in which one of the pistons is fitted may be so constructed as to allow of its plane being changed; and it will be found that in any position, the pressure, or additional pressure, upon the piston is the same.

8. If a mass of fluid be at rest, any portion of it may be contemplated as a separate body surrounded by fluid, which presses upon its surface perpendicularly at all points.

It follows therefore, from the laws of statics, that the resultant of the fluid pressures upon the portion considered is equal and opposite to the resultant of the extraneous forces, such as gravity or other attractive forces, which are in action upon that portion.

9. The two principles of the equal transmission of pressure and of the equality of pressure in all directions, for the truth of which we have appealed to experience, can be deduced from the fundamental property of a fluid, stated as an axiom in Art. (2).

* If A and B be two pistons of any shape and size, they can be divided into small areas of the same shape and size, and by making these areas small enough, it will be seen that their numbers will be ultimately in the ratio of the areas A and B.

10.   *The equality of pressure at any point in all directions.*

We shall prove this for the case of fluids at rest under the action of gravity, that is, for heavy fluids at rest.

Take a small rectangular wedge or prism of fluid, having its sides horizontal and vertical, and its plane ends vertical, and let ABC be its section by a vertical plane bisecting its length. This prism is at rest under the action of gravity and of the pressures of the fluid on its ends and sides. The ends are supposed to be perpendicular to the sides of the prism; hence, the pressures on these ends being perpendicular to all the other forces must balance each other, and the pressures on the sides AC, CB, BA, must balance the weight.

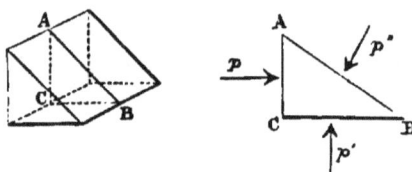

Taking d for the length of the wedge, a, b, c the sides of the triangle, w for the weight of an unit of volume of the fluid, and p, p', p'' for the measures of the pressures on the sides AC, CB, BA, these pressures are

$$pbd,\ p'ad,\ \text{and}\ p''dc,$$

and the weight is $\frac{1}{2}abdw$.

Hence resolving vertically and horizontally,

$$\tfrac{1}{2}abw = p'a - p''c \cos B,$$
$$pb = p''c \sin B;$$

but $a = c \cos B$, and $b = c \sin B$;

$$\therefore\ p = p'',\ \text{and}\ p' - p'' = \tfrac{1}{2}bw.$$

If now we suppose the sides a, b indefinitely diminished, in which case p, p' and p'' will be the pressures in different directions at the point C, we shall have $p' = p''$, and therefore the three pressures are equal *.

By turning the wedge round AC and changing the angle A and B it will be seen that the proposition is true for all directions.

11.   *The transmission of pressure.*

Let A and B be two points in a fluid at rest, and about the straight line AB as axis describe a cylinder having plane ends perpendicular to AB.

---

* In strictness $p'pp''$ are the measures of the mean pressures on the sides of the wedge, but a reference to Art. 4 on the measure of variable pressure will shew why it is unnecessary to repeat an explanation already made.

The equilibrium of the cylinder is maintained by the fluid pressures on its ends, which are parallel to its axis, by the fluid pressures on its curved surface, which are perpendicular to its axis, and by its weight.

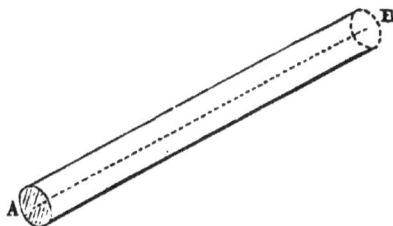

Now resolving along AB, the difference of the pressures at A and B must be equal to the resolved part of the weight in the direction BA, and the weight remaining the same, any change of pressure at A involves the same change at B. Moreover, if fluid be contained in a vessel of any shape, and the straight line AB do not lie entirely in the fluid, the two points may be connected by a series of straight lines such as $ACDB$, and any change of pressure at A produces an equal change at C, and therefore, taking account of the previous article, the same change is produced at D, and therefore at B.

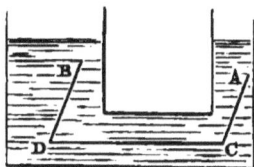

12. *The Hydrostatic Bellows* is a machine illustrating the principle of the transmission of fluid pressure.

B is the top of a cylinder having its sides made of leather, and CA is a pipe leading into it. If this vessel and the pipe be filled with water and a pressure applied at A, a very great weight upon B may be raised by a small pressure at A, the weight lifted being greater in proportion to the size of B.

Even without water weights may be raised by simply blowing into the tube A.

13. *The Hydrostatic Paradox.*

Any quantity of liquid, however small, may be made to support any weight, however large.

This is another mode of enunciating the same principle. For in the previous figure we may suppose the tube CA

extended vertically, and the pressure produced by pouring in water to a considerable height, so as to produce a pressure at A by means of the column of liquid above it. The tube may be very thin, so that the pressure upon the section A of the tube may be very small, but, as this pressure is transmitted to every portion of the surface B, which is equal to the section A, the force produced can be as large as we please. To increase the upward force on B we must enlarge the surface B, or increase the height of the column of liquid in the tube, and the only limitation to the increase of the force will be the want of sufficient strength in the pipe and cylinder to resist the increased pressure. By making the height BC very small, and the tube A of very small bore, the quantity of liquid can be made as small as we please, and hence the paradoxical statement made above.

Hydraulic Presses.

14. The transmission of fluid pressure is the principle upon which Hydraulic or Hydrostatic Presses are constructed.

Thus, if A, B be two pistons working in hollow cylinders connected by a pipe C, and filled with water, any force applied to the piston B is transmitted to A, and the force upon A is greater than the force on B in the ratio of the area of A to B.

This is a Hydraulic Press in its simplest form. Practically it is requisite to have a reservoir from which more water can be obtained by a pump, and we therefore defer the description of a complete Hydrostatic Press until the principle of the Pump has been explained.

The Safety-Valve.

15. In many machines, and especially in steam engines, a very great fluid pressure may be produced, and the strength of the machine may be very severely tried: in order to guard against accidents arising from the bursting of the machine a *safety-valve* is employed, which serves to indicate the existence of too large a pressure.

Various forms may be used, but the principle of the safety-valve is simply that of the uniform transmission of pressure in a fluid.

Thus if BC be one of the connecting tubes through which the fluid passes, and D a small tube opening out of BC, the pressure on a lid at the end of D will measure the fluid pressure within, and if the lid be of a suitable weight, it will be lifted when the pressure is greater than the machine is intended to bear. Suppose, for instance, the greatest permissible pressure of the fluid to be 500 lbs. on a square inch, and the sectional area of the tube D to be $\frac{1}{16}$th of a square inch, then a weight of $\frac{500}{16}$ or $31\frac{1}{4}$ lbs. will be lifted when the pressure exceeds 500 lbs. The weight employed may be diminished if the lid be moveable about a hinge at A, and a weight w be placed at some little distance from A.

Ex. 1. The cross section of the tube D is a square, the side of which is one-fourth of an inch; the lid is moveable about the end A, and a 5 lbs. weight is attached to the lid at the distance of two inches from the hinge.

In this case the resultant fluid pressure will act at the centre of the square, and if the greatest fluid pressure on a square inch is equal to the weight of x lbs., we have a force equal to the weight of $\frac{x}{16}$ lbs. counterbalanced by the weight of 5 lbs.;

$$\therefore \quad \frac{x}{16} \times \frac{1}{8} = 5 \times 2, \text{ or } x = 1280.$$

Ex. 2. Taking the same square tube, but taking the distance $A W$ to be $2\frac{1}{2}$ inches, find the weight which will indicate a pressure on a square inch equal to the weight of 3200 lbs.

16. It will be seen that in Hydrostatic presses, as in all machines, the principle holds that *what is gained in power is lost in motion.*

Thus, if there be two apertures in a closed vessel, fig. Art. 5, and the piston B be forced down through any given space, the piston A is forced upwards if the fluid be incompressible, through a space which is less as the area of A is greater.

This is a simple case of the principle of virtual work which we proceed to demonstrate, as applied to incompressible fluids.

Let A, B, C,... be the areas of a number of pistons working in cylindrical pipes fitted into the sides of a closed vessel which is filled with fluid. Let the pistons be moved in any manner so that the fluid remains in contact with them, and a, b, c,... be the spaces through which they are moved, these quantities being positive or negative, as the pistons are pushed inwards or forced outwards.

Then, since the volume of fluid is the same as before, it follows that

$$Aa + Bb + Cc + \ldots = 0,$$

the positive portions, that is, the volumes forced in, being balanced by the negative portions, or the volumes forced out.

But if P, Q, R,... be the forces on each piston,

$$P : Q : R : \ldots = A : B : C : \ldots$$
$$\therefore Pa + Qb + Rc + \ldots = 0;$$

or the sum of the products of each force into the space through which its point of application is moved is equal to zero; and observing that a, b, c,... are proportional to infinitesimal displacements of the pistons, this is the equation of virtual work.

17. It is not to be imagined that there exists any substance in nature exactly fulfilling the definition which has been given of a fluid. Just as the ideas of a perfectly smooth surface and a perfectly rigid body are formed from observations of bodies of different degrees of rigidity, and surfaces of different degrees of smoothness, so the idea of

perfect fluidity is suggested. Nevertheless in the cases of fluids at rest the theoretical properties of fluids derived from this definition will be found to agree with facts, and it is in cases of fluid motion that sensible discrepancies will be found. Thus, a cup of tea set rotating will gradually come to rest, proving the existence of a friction between the liquid and the tea-cup, and also between the particles of the liquid, since the dragging force is gradually communicated from the outer to the inner portions. The motion of water in inclined tubes also indicates the existence of a frictional action amongst the particles of water.

18. Recognizing the fact that all fluids possess, more or less, the characteristic of viscosity, we can give a definition which will include fluids of all degrees of viscosity.

A fluid is an aggregation of particles which yield to the slightest effort made to separate them from each other, if it be continued long enough.

It follows from this definition that in a fluid in equilibrium there can be no tangential action, or shearing stress, and therefore that the pressure on any surface in contact with the fluid is normal to that surface.

Hence all theorems relating to the equilibrium of fluids are true for fluids of any degree of viscosity.

EXAMINATION UPON CHAPTER I.

1. DISTINGUISH between compressible and incompressible fluids. Are any liquids absolutely incompressible ?

2. State the property which is assumed as the basis of all reasonings upon fluid action.

3. Define the measure of fluid pressure.

4. It is found that the pressure is uniform over the whole of a square yard of a plane area in contact with fluid, and that the pressure on the area is equal to the weight of 13608 lbs.; find the measure of the pressure at any point, 1st, when the unit of length is an inch, 2nd, when it is two inches.

5. The plane of a rectangle, in contact with fluid, is vertical, two of its sides are horizontal, and it is known that at all points of the same horizontal line the pressure is the same. The pressure on the rectangle, for all values of h, is $wbh (a+h)$ where b is the width and h the height of the rectangle; find the pressure at any point of the upper side. (Art. 4.)

6. A cylindrical pipe which is filled with water opens into another pipe the diameter of which is three times its own diameter: if a force of 20 lbs. wt. be applied to the water in the smaller pipe, find the force on the open end of the larger pipe, which is necessary to keep the water at rest.

7. Account for the fact of the transmission of pressure through a liquid.

Mention any direct practical application of this principle.

8. In a Hydrostatic Bellows (Art. 12), the tube A is $\frac{1}{8}$th of an inch in diameter, and the area B is a circle, the diameter of which is a yard. Find the weight which can be supported by a pressure of 1 lb. on the water in A.

9. A safety-valve consists of a heavy rectangular lid which is horizontal when it closes the aperture beneath it, and is moveable about one side. The aperture being a square which has one side coincident with the fixed side of the lid, find the maximum pressure marked by the valve.

10. If the side of the aperture in the preceding question is a quarter of an inch and if the side of the lid is two and a half inches, find what must be the weight of the lid in order that a maximum pressure equal to the weight of 800 lbs. may be indicated.

11. Prove the principle of virtual work in the case of the sixth question.

12. A triangular area ABC is exposed to fluid pressure, and it is found that if any straight line PQ be drawn parallel to BC, and at a distance x from A, the pressure on the area APQ is px^2; find the pressure at A, and also at any point of the line BC.

13. A plane area is exposed to fluid pressure, and it is found that the pressure on any circular portion of the area, having its centre at a fixed point, is proportional to the cube of its radius; prove that the pressure at any point of the area is proportional to its distance from the fixed point.

14. A strong cylindrical tube, one foot in diameter inside, and ten feet in length, is filled with distilled water, and closed with a piston to which a pressure of 10000 lbs. is applied; shew that the resulting compression of the water will be nearly $\frac{1}{24}$th of an inch.

15. Let the tube in the preceding question be closed with the exception of a circular aperture one inch in radius, and let there be fitted to this aperture a straight tube of the same radius.

If a length of ten feet of this smaller tube is filled with water, and if a pressure equal to the weight of 58π lbs. is applied by a piston to the outer surface of the water, prove that the piston will be forced in through ·87024 inches.

CHAPTER II.

19. In the classification of fluids the most prominent division is between gases and liquids, or compressible and incompressible fluids, as they are sometimes termed, and under these two heads all fluids are naturally ranged. It has been remarked already that the compressibility of liquids is practically insensible, and for all ordinary purposes unimportant.

It will be found, however, that the theory of sound is partly dependent on this compressibility, and it is therefore of importance at once to recognize its existence.

There are many other characteristics which distinguish fluids from each other, such as colour, degree of transparency, chemical qualities, viscosity, &c., but in the theory of Hydrostatics the characteristics which it is especially necessary to consider are *density* and *weight*.

Density.

20. DEFINITION. *The density of any uniform substance is the mass of an unit of volume of the substance.*

The mass of a body is the quantity of matter contained in it, measured in terms of some standard unit.

In this country the standard unit of mass is taken to be the quantity of matter contained in a *Pound Avoirdupois.*

The standard Pound Avoirdupois is really a lump of platinum, which is kept under a glass case in certain offices in London, and from which copies can be made. The mass of a body is therefore the number of pounds avoirdupois which it contains, and the density of a substance is the number of pounds in an unit of volume.

For example it is known that the mass of a cubic foot of water is 1000 oz., or 62·5 lbs., and it follows that, if we take a foot as the unit of length, the density of water is 62·5.

Again it is known that the density of mercury is 13·568 times that of water; the density of mercury is therefore 848.

Under ordinary conditions of pressure and temperature, the density of atmospheric air is to that of water in the ratio of ·0013 : 1, and therefore the mass of a cubic foot of air is ·08125 lb., or 1·3 oz., or 568·75 gr. It is to be borne in mind that the pound avoirdupois contains sixteen ounces, and also that it contains seven thousand grains.

21. If M is the mass of a volume V of uniform substance, the density of which is ρ, then, in accordance with the definition, we have the equation

$$M = \rho V.$$

It will be seen that the measure of the mass of a body is an absolutely fixed quantity, and is not in any way dependent upon time or place.

22. In France, and on the Continent generally, the metric system of units is adopted. In this system the unit of mass is the Kilogramme, which represents the mass of one Litre of water, that is, one thousand cubic centimetres of water, at its maximum density.

One kilogramme is 1000 grammes, so that a gramme is the mass of a cubic centimetre of water at its maximum density.

The Pound Avoirdupois is 453·59265 grammes, so that one kilogramme is, very nearly, 2·2 lbs.

The Pound Avoirdupois contains 7000 grains; one grain is therefore very nearly ·0648 grammes, or 64·8 milligrammes.

Again, the metre being 39·370432 inches, or 3·280869 feet, one foot contains 30·4797 centimetres, and one cubic inch contains 16·387 cubic centimetres. Hence it follows that the litre is very nearly ·0353 of a cubic foot, or 61 cubic inches.

It may be useful to place some of the relations between the English and French measures of length and mass in a tabular form.

Length, Area and Volume.

1 metre	=	39·370432 inches.
1 centimetre	=	·393704 „
1 inch	=	2·5400 cm.
1 foot	=	30·4797 cm.
1 square inch	=	6·4516 sq. cm.
1 square foot	=	929·01 sq. cm.
1 cubic inch	=	16·387 cub. cm.
1 cubic foot	=	28316 cub. cm.

Mass.

1 kilogramme	= 2·2046212 lbs.
	= 15432·3484 grains.
1 grain	= ·064799 grammes
1 oz. avoirdupois	= 28·34954 „
1 lb. „	= 453·59265 „

Weight.

23. *Weight and Intrinsic Weight.*

The weight of a body is the force exerted upon it by the action of gravity.

If a body, hanging freely, is supported by a single string, the tension of the string is equal to its weight.

Thus, if a mass of one pound is held in the hand, or supported by a string, the upward-force exerted by the hand, or the tension of the string, is equal to the weight of one pound.

Now, although the mass of a pound is an invariable quantity, the weight of the pound is a variable force, being both local and temporary; it is different at different places, and, at the same place, it varies from time to time.

If a given mass, at a given place, say London for instance, be suspended from a spring, it will be seen to stretch the spring to a certain extent.

If the mass be carried to a place nearer the equator than London, the spring will be less extended, but if it be carried northwards, the spring will be more extended.

In the former case the weight will be less, and in the latter case greater than it is in London.

24. DEFINITION. *The intrinsic weight of a substance is the weight of an unit of volume of the substance, expressed in terms of some standard unit of weight.*

Hence if w represents the intrinsic weight of a substance, and if W be the weight of the volume V of the substance, we have the equation

$$W = wV.$$

If the standard unit of weight is taken to be the weight of a pound at a particular place, then, at that place, the numerical values of ρ and w will be the same.

Practically the difference due to change of locality is very slight, the ratio of polar to equatoreal gravity being

$$32{\cdot}2527 : 32{\cdot}088.$$

When we say that ρ is the density of a substance, we assert that the volume V of it contains ρV units of mass.

When we say that w is the intrinsic weight of a substance we assert that the action of gravity upon a volume V of it is equivalent to wV units of force.

When we say that a portion of some substance weighs x pounds, or y kilogrammes, we assert that the action of gravity upon it is equivalent to the action of gravity upon x pounds, or upon y kilogrammes.

In measuring the pressure of fluids upon surfaces, we shall generally take the weight of a pound as the unit of force.

25. In the previous articles we have considered homogeneous bodies only.

If the density be variable, and if it vary continuously from point to point, we can determine the density at any point by taking a small volume v of the substance containing the point, and finding the mass m of this volume. The expression m/v will be the mean density of the volume v, and the ultimate value of m/v, when v and therefore m, are indefinitely diminished, will be the measure of the density at the point.

26. In order to render more clear the mathematical conception of a continuously varying substance, imagine a number of homogeneous strata of equal thickness t placed on each other, and suppose the density of the lowest stratum to be ρ and of the highest ρ', and of the intermediate strata let the densities increase by successive additions from ρ' to ρ.

If now we suppose the thickness of each stratum t to become indefinitely small, and the number of intermediate strata to become indefinitely large, while the densities of the extreme strata ρ, ρ' remain the same, the densities of the intermediate strata which are to increase from ρ' to ρ will differ from each other by infinitely small quantities, and we can thus form an idea of a continuously varying medium.

This mode of viewing continuity by means of discontinuity is necessary for the purposes of mathematical calculation.

The atmosphere in a state of rest is a case in point, as its density decreases continually as the height increases.

27. The density of a mixture may be determined by the previous formula $M = \rho V$.

Thus, if volumes V, V', V'',... of fluids whose densities are ρ, ρ', ρ''... be mixed together, and if the mixture form a homogeneous mass, and no change of volume occur from chemical action, the whole mass

$$= \rho V + \rho' V' + \rho'' V'' + \dots = \Sigma (\rho V),$$

and the whole volume $= V + V' + V'' + \dots = \Sigma (V)$;

$\therefore$ the density of the mixture $= \dfrac{\Sigma (\rho V)}{\Sigma (V)}$.

Specific Gravity.

28. DEFINITION. *The specific gravity of a substance is the ratio of the weight of any volume of the substance to the weight of an equal volume of a standard substance.*

In other words the specific gravity is the ratio of the density of the substance to the density of the standard substance.

Hence, if s is the specific gravity of a substance, and if W is the weight of a volume V of the substance, and w the intrinsic weight of the standard substance we have the equation

$$W = wsV.$$

Distilled water, at the temperature $4°$ C., is generally taken as the standard substance, and in that case the weight, in lbs. weight at the place, of the volume V cubic feet is given by the equation

$$W = 62{\cdot}5sV.$$

It will be seen that the ratio of the densities of two different substances is the same as the ratio of their intrinsic weights, and is also the same as the ratio of their specific gravities.

29. *To find the specific gravity of a mixture of given volumes of any number of fluids, whose specific gravities are given.*

Let V, V', V''... be the volumes of fluids of which the specific gravities are s, s' s'',...

Then the weight of the mixture is

$$62{\cdot}5 \{sV + s'V' + s''V'' + \ldots\} \text{ or } 62{\cdot}5 \, \Sigma \, (sV),$$

and therefore if σ be the specific gravity of the mixture,

$$62{\cdot}5 \, \sigma \, . \, \Sigma \, (V) = 62{\cdot}5 \, \Sigma \, (sV),$$

or

$$\sigma = \Sigma \, (sV) \div \Sigma \, (V).$$

If by any chemical action the volume becomes U instead of $\Sigma (V)$, we shall then have

$$\sigma U = \Sigma \, (sV).$$

30. *To find the specific gravity of a mixture when the weights and the specific gravities of the components are given.*

If W, W',... are the weights, and s, s',... the specific gravities, the volumes are

$$\frac{1}{62{\cdot}5} \frac{W}{s} , \qquad \frac{1}{62{\cdot}5} \frac{W'}{s'} , \ldots$$

Hence, if σ is the specific gravity of the mixture.

$$62\cdot5 \cdot \sigma \cdot \Sigma\left(\frac{1}{62\cdot5} \, \frac{W}{s}\right) = \Sigma \cdot (W),$$

or
$$\sigma \cdot \Sigma\left(\frac{W}{s}\right) = \Sigma(W).$$

31. The practical methods of determining the specific gravities of solids, liquids, and gases will be discussed in a future chapter.

For solids and liquids tables of specific gravity are usually given with reference to distilled water at its maximum density as the standard.

Gases and vapours are, however, generally referred to atmospheric air at the same temperature and under the same pressure as the gases themselves.

EXAMINATION UPON CHAPTER II.

1. Distinguish between the measures of mass, density, and specific gravity.

2. Find the masses in lbs., of a cubic yard and a cubic inch of water, and also of a cubic yard and a cubic inch of mercury.

3. Find the number of grammes in a cubic foot of water, and in a cubic foot of mercury.

4. Find what fraction of an ounce is the mass of a cubic centimetre of water.

5. If the mass of 10 cubic centimetres of a liquid is one gramme, what is its density in pounds per cubic foot?

6. The specific gravity of cork being ·24, find what volume of water weighs as much as a cubic yard of cork.

7. Find the specific gravity of an alloy of gold and copper in the ratio of 11 : 1, the specific gravities being 19·4 and 8·84.

8. Equal volumes of two fluids whose specific gravities are 5 and 7 are mixed together ; find the specific gravity of the mixture.

If equal weights of the same fluids are mixed together, find the specific gravity of the mixture.

9. If a cubic inch of a standard substance weigh ·45 of a lb., what is the weight of a cubic yard of a substance of which the density is 5 ?

10. Equal weights of two fluids, of which the densities are ρ and 2ρ, are mixed together, and one-third of the whole volume is lost; find the density of the resulting fluid.

11. Taking water as the standard, find the weight of a cubic yard of a substance of which the specific gravity is ·12.

12. A cubic inch of a substance weighs $1\frac{9}{3}\frac{2}{4}\frac{3}{5}\frac{3}{8}$ths of a lb.; find its specific gravity referred to water.

13. A mixture is formed of equal volumes of three fluids; the densities of two are given and the density of the mixture is given; find the density of the third fluid.

14. Volumes V, V'' of two fluids, the specific gravities of which are σ, σ', are mixed together, and the specific gravity of the mixture is s; find the change in volume.

EXAMPLES.

1. A mixture is formed of two fluids; the density ρ of the mixture, the ratio, $m : 1$ of the volumes, and the ratio, $n : 1$ of the densities are given; find the densities of the fluids.

2. Two fluids of equal volume and of densities ρ, 2ρ, lose one-fourth of their whole volume when mixed together; find the density of the mixture.

3. A mixture is formed of equal volumes of n fluids, the densities of which are in the ratio of the numbers $1, 2, 3,...n$; find the density of the mixture. Also find the density of the mixture when the volumes are in the ratio:—1st, of the numbers $1, 2, 3,...n$, and 2nd, of the numbers $n, n-1,...3, 2, 1$.

4. Having given the specific gravity σ of a mixture formed of equal volumes of two fluids, and also the specific gravity σ' of a mixture formed by taking a volume of one fluid double that of the other, find the specific gravities of the fluids.

5. When a vessel is filled by means of equal volumes of two fluids, the specific gravity of the compound is $\frac{4}{3}$ of what it would have been if the vessel had been filled by means of equal weights of the fluids. Compare the specific gravities of the two fluids.

6. If the true specific gravity of milk be 1·031, what quantity of water must be mixed with 10 gallons of milk to reduce its specific gravity to 1·021?

7. If the centimetre be the unit of length, and if a centimetre cube of water be taken as the unit of mass and called a gramme, prove that the number of grammes in the earth is $6·15 \times 10^{27}$, the diameter of the earth being taken to be $1·275 \times 10^9$ centimetres and the mean density 5·67.

8. If the weight of 28 grains is taken as the unit of weight, what must be the unit of length in order that the numerical measure of the weight of a body may be equal to the product of its volume and its specific gravity?

9. The mixture of a gallon of A with λ_1 lbs. of B has a specific gravity σ_1, with λ_2 lbs. of B a specific gravity σ_2, with λ_3 lbs. of B a specific gravity σ_3; find the specific gravities of A and B.

10. Two liquids are mixed together, first by weights in the proportion of their volumes of equal weights, and secondly by volumes in the proportion of their weights of equal volumes; compare the specific gravities of the two mixtures.

CHAPTER III.

32. *THE pressure of a liquid at rest is the same at all points of the same horizontal plane.*

Take a thin cylindrical portion AB of the liquid, having its axis horizontal, and its ends A, B vertical, and consider the equilibrium of this portion. We have then a body AB kept at rest by the fluid pressures on its curved surface, all of which are perpendicular to the axis of the cylinder, by the pressures on the two ends, which are horizontal, and by the weight of the body.

If p and p' be the measures of the pressures at A and B, and α the area of each end, which is taken to be very small in order that the pressure may be sensibly uniform over the whole of either end, the pressures on the ends are $p\alpha$ and $p'\alpha$, and since these balance each other we have

$$p = p'.$$

This proof also holds good for the case of gases, and for heterogeneous liquids.

33. *To find the pressure at any given depth in a heavy homogeneous liquid at rest.*

Taking any point P in the fluid, draw PA vertically to the surface, and describe a thin cylinder about PA with its base horizontal.

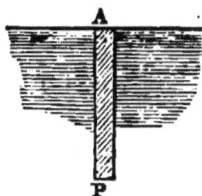

Then the portion of fluid PA is kept at rest by the fluid pressure on the end P, its weight, and the fluid pressures on the curved surface, which are all horizontal.

Hence the fluid pressure on P must be equal to the weight, and therefore, if α be the area of the base, w the weight of an unit of volume, and p the pressure at P,

$$p\alpha = w\alpha . AP,$$

or
$$p = w . AP;$$

that is, the pressure at any depth varies as the depth below the surface.

Similarly, if P and Q be any two points in the same vertical line, it will be seen, by describing a cylinder PQ, that the difference of the pressures on the ends P and Q of the cylinder must be equal to the weight of the cylinder.

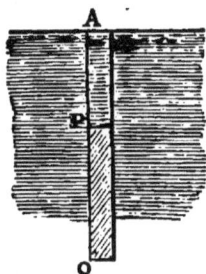

Hence if p, p' be the pressures at P and Q,

$$p'\alpha - p\alpha = w\alpha . PQ,$$

or
$$p' - p = w . PQ;$$

that is, *the difference of the pressures at any two points varies as the vertical distance between the points.*

34. Let the cylinder of which AP is the axis be bounded at P by a plane not horizontal, and let a' be its area, and θ its inclination to the horizon.

Then for the equilibrium of the cylinder, taking p' as the pressure at P upon a', we have by resolving vertically,

$$p'a' \cos \theta = wa.AP,$$

but
$$a = a' \cos \theta;$$

$\therefore p' = w . AP$, which is independent of θ.

We thus have another proof of the proposition that the pressure at any point is the same in all directions.

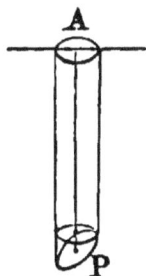

It may be perhaps objected to the proof of Art. 33 that the surface at A is assumed to be horizontal. By making the cylinder AP a very thin cylinder, that is, of very small radius, it will be seen that its weight is sensibly $waAP$, and therefore that the proof does not depend on any assumption as to the form of the surface.

Or, to reason more strictly, draw two horizontal planes through the highest and lowest points B, A of the small portion AB of the surface intercepted by the cylinder.

Then, if the radius of the cylinder be indefinitely diminished, these two planes will coalesce.

If z and z' be the heights above P of these planes, the weight of the cylinder lies between

$$waz \text{ and } waz',$$

and therefore p lies between

$$wz \text{ and } wz',$$

and ultimately when the planes coalesce,

$$p = wz.$$

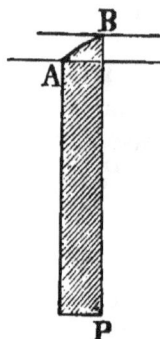

35. *Difference of pressures at any two levels in an elastic fluid.*

We have already mentioned in Art. 20, that gases are heavy bodies; hence, by the same reasoning as in Art. (33), if P and Q be two units of area in an elastic fluid, P being vertically above Q, the difference of the pressures at P and Q is equal to the weight of the column of fluid PQ. This column is not of uniform density, and hence the law of variation of the pressure at different levels in an elastic fluid does not present itself in a simple form. Further information will be found in Chapter V.; at this point we need only call attention to the fact that the pressure decreases as we ascend in an elastic fluid.

36. *The surface of a liquid at rest is a horizontal plane.*

Take two points P, Q, in the same horizontal plane, within the liquid, and draw PA, QB vertically to the surface.

Then pressure at $P = w \cdot AP$,

pressure at $Q = w \cdot BQ$,

and these are equal; therefore AP and BQ are equal, and A, B are in the same horizontal plane. Similarly

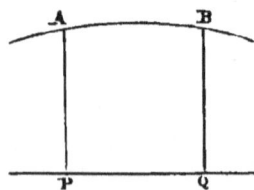

any other point in the surface can be proved to be in the same horizontal plane with A or B.

Or we might have argued that, since the pressures are equal at all points of the same horizontal plane, conversely, all points at which the pressures are equal are in the same horizontal plane, and therefore all points in the surface, at which the pressure is either zero, or equal to the atmospheric pressure, must be in the same horizontal plane.

37.   The pressure of the atmosphere is found to be about 14·73 lbs. to a square inch, or very nearly 15 lbs. We can hence calculate the pressure upon any given area, and, if Π be the atmospheric pressure on the unit of area, the pressure at a depth z of a fluid, the surface of which is exposed to atmospheric pressure, will be

$$wz + \Pi.$$

38.   *Illustration.*   Take a hollow glass cylinder open at both ends; in contact with the lower end, and closing that end, place a heavy flat disc supported by a string passing up the cylinder.

Holding the string, depress the cylinder in a vessel of water, and it will be found that, at a certain depth, the string may be loosened, and the disc will remain in contact with the cylinder, being supported by the pressure of the water beneath.

If W be the weight of the disc and r the radius of the cylinder, the requisite depth (x) of the disc is given by the equation

$$W = wx\pi r^2.$$

The presence or absence of the atmosphere will not affect this depth, since the pressure of the atmosphere downwards on the disc would be counteracted by the pressure upwards, transmitted from the surface of the water.

39.   If in Art. (32) the line AB do not lie entirely within the fluid, we can still prove the truth of the proposition by the aid of Art. (33).

For A and B can be connected by horizontal and vertical lines as AC, CD, DB, and

pressure at B

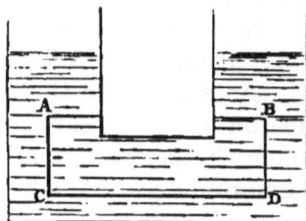

$$= \text{pressure at } D - w \,.\, BD$$

$$= \text{pressure at } C - w \,.\, AC$$

$$= \text{pressure at } A.$$

40. Hence it appears that all points on the surface of a liquid, at which the pressure is either zero or is equal to the constant atmospheric pressure, must be in the same horizontal plane, and that this is true even though the continuity of the surface be interrupted by the immersion of solid bodies, or in any other way.

This sometimes appears under the form of the assertion that *liquids maintain their level*, and an experimental illustration may be employed as in the figure.

A number of glass vessels of different forms, all open into a closed tube or vessel AB, and it is found that if water be poured into any óne of the tubes, it will, after filling the tube AB, rise to exactly the same vertical height in every one of the tubes, and if any portion be withdrawn from any of the vessels, that the water will sink to its new position of rest through the same vertical height in each.

An important practical illustration of this principle is seen in the construction by which towns are supplied with water. A reservoir is placed on a height, and pipes leading from it carry the water to the tops of houses or to any point which is not higher than the surface of the water in the reservoir, and these pipes may be carried under ground or over a road, provided that no portion of a pipe is above the original level.

41. *The common surface of two liquids that do not mix is a horizontal plane.*

Take two points P, Q in the lower fluid, both in the same horizontal plane, and let vertical lines PA, QB to the surface of the upper fluid meet the common surface of the fluids in C and D.

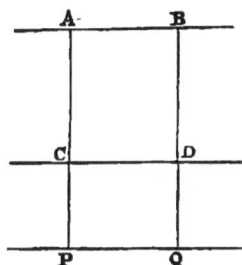

Then, if w' be the intrinsic weight of the lower fluid, and w of the upper,

pressure at $P = w' . CP +$ pressure at C

$$= w' . CP + w . CA,$$

and pressure at $Q = w' . QD + w . DB$;

$$\therefore \ w' . CP + w . CA = w' . QD + w . DB.$$

Also AB is horizontal, and therefore

$$CP + CA = QD + DB;$$

$\therefore$ multiplying by w and subtracting,

$$(w' - w) . CP = (w' - w) \, QD,$$

or $CP = QD$, and therefore CD is horizontal.

42. *If two liquids that do not mix together meet in a bent tube, the heights of their upper surfaces above their common surface will be inversely proportional to their intrinsic weights.*

Let A and B be the two surfaces, C the common surface, and w, w', the intrinsic weights of the liquids.

Let horizontal planes through A, B, and C meet a vertical line in a, b, and c, and take C' in the denser fluid in the same horizontal plane as C.

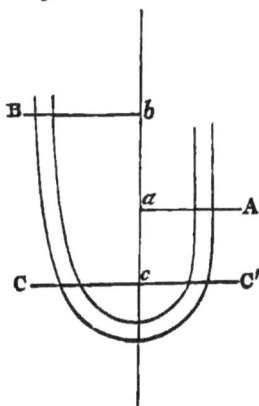

The pressure at $C = w \cdot bc$, and the pressure at $C' = w \cdot ac$, and these are equal, by Art. 32;

$$\therefore \ w \cdot bc = w' \cdot ac,$$

or
$$bc : ac = \frac{1}{w} : \frac{1}{w'}.$$

Since the densities are in the ratio of the intrinsic weights, it follows that the heights of the upper surfaces above the common surface are inversely as the densities of the fluids.

43. *Two fluids that do not mix are contained in the same vessel; it is required to find the pressure at a given depth in the lower fluid.*

Let P be the point in the lower fluid, PBA a vertical line meeting the common surface in B. Describe a small cylinder about AP, and consider the equilibrium of the fluid within it.

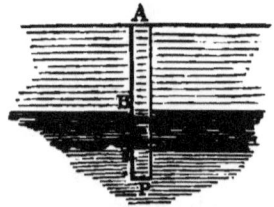

Then, if p be the pressure at P and α the sectional area of the cylinder,

$p\alpha = \text{weight of } ABP = w\alpha \cdot AB + w'\alpha \cdot BP,$

w and w' being the intrinsic weights,

or
$$p = w \cdot AB + w' \cdot BP.$$

This might have been at once inferred from the equation

$$p = w' \cdot BP + \text{pressure at } B,$$

for the pressure at $B = w \cdot AB$.

And in the same manner the pressure at any point of a mass of fluid containing any number of strata of different densities can be determined.

If the surface A be subject to the atmospheric pressure Π,

the pressure at $P = w' \cdot BP + w \cdot AB + \Pi$.

44. We now proceed to consider two simple cases of the pressure of a fluid on plane surfaces.

PROP. *The pressure of a liquid on any horizontal area is equal to the weight of a column of the liquid of which the area is the base and of which the height is equal to the depth of the area below the surface.*

For, if z be the depth, the pressure at every point is wz ;

$\therefore$ if κ be the area, the pressure upon it $= wz\kappa$,

and $z\kappa$ is the volume of the column described.

It will be seen that this is independent of the form of the vessel containing the fluid.

This result may also be obtained in the following manner.

Draw through the boundary of K vertical lines to the surface, and consider the equilibrium of the portion of fluid

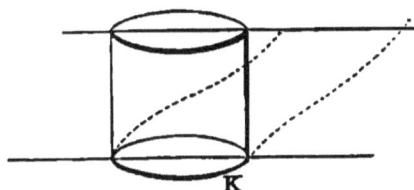

enclosed. The pressure of the surrounding fluid is entirely horizontal, and therefore the pressure on the base must be equal to the weight of the fluid enclosed.

If the vessel be of the form indicated by the dotted line so that the actual surface does not extend over the area K, we may suppose the fluid extended over K by enlarging the vessel, and the pressure at any point of K will not be changed. Hence the above reasoning is applicable to this case also.

Thus if a hollow cone, vertex upwards, be just filled with water, and if r be the radius of the base and h the height of the cone, the pressure on the base $= w\pi r^2 h$, that is, the weight of the cylinder of fluid on the same base as the cone, and of the same height.

45. *A plane area in the form of a rectangle is just immersed in liquid with one edge in the surface, and its plane inclined at an angle θ to the vertical; it is required to find the pressure upon it.*

Let the figure be a vertical section perpendicular to the side b of the rectangle in the surface, $AB\ (=a)$ being the section of the rectangle, and draw a vertical plane BC through the lower side B. Then the weight of the fluid in ABC is supported by the plane AB, since the pressure on BC is horizontal.

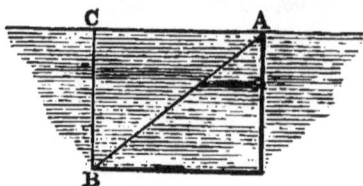

Hence if R be the pressure on AB, perpendicular to its plane,

$$R \sin \theta = \text{weight of } ABC = \tfrac{1}{2}w \,.\, AC\,.\,BC\,.\,b$$

$$= \tfrac{1}{2}wba^2 \sin \theta \cos \theta \,;$$

$$\therefore R = \tfrac{1}{2}wba^2 \cos \theta = wba\,.\,\tfrac{1}{2}a \cos \theta,$$

that is, the pressure is the weight of a column of fluid of which the rectangle is the base, and the height is equal to the depth of the middle point of AB below the surface.

Since the direction of R makes an angle θ with the horizon, it follows that the horizontal component of R is

$$\tfrac{1}{2}wba^2 \cos^2 \theta.$$

Now the fluid in ABC is kept at rest by the horizontal pressure on BC, by its weight, and by the reaction R.

Hence the pressure on $BC = R \cos \theta = \tfrac{1}{2}wba^2 \cos^2 \theta$

$$= w\,.\,ba \cos \theta \,.\, \tfrac{1}{2}a \cos \theta$$

$$= w\,.\,(\text{area } BC)\,(\text{depth of middle point of } BC),$$

the same law as for AB.

This also appears from the value of R by putting $\theta = 0$.

The results thus obtained are generalized in the following article in which a different method is adopted.

Whole Pressure.

46. DEF. *The whole pressure of a fluid on any surface is the sum of all the normal pressures exerted by the fluid on every portion of the surface.*

In the case of a plane, the pressure at every point is in the same direction and the whole pressure is the same as the resultant pressure. In the case of curved surfaces, the whole pressure is merely the arithmetical sum of all the pressures acting in various directions over the surface.

PROP. *The whole pressure of a liquid on a surface is equal to the weight of a column of liquid of which the base is equal to the area of the surface, and the height is equal to the depth of its centroid below the surface of the liquid.*

Let the surface be divided into a great number of very small areas a_1, a_2, a_3,... and let z_1, z_2, z_3... be the depths below the surface of the centroids of these areas. If the areas be taken very small, each may be considered plane, and the pressures upon them will be respectively

$$wa_1z_1, \quad wa_2z_2,...$$

taking the pressure over each area to be uniform.

Hence the whole pressure $= w\Sigma(az)$.

But, if $\bar{z}$ be depth of the centroid of the surface,

$$\bar{z} = \frac{\Sigma(az)}{\Sigma(a)} * ;$$

∴ whole pressure $= w\bar{z}\Sigma(a)$

$= w\bar{z}S$, if S be the area of the surface,

and $\bar{z}S$ is the volume of the column described.

Ex. 1. A rectangle is immersed with two sides horizontal, the upper one at a given depth (c), and its plane inclined at a given angle (θ) to the vertical.

Let a be the horizontal side, b the other side.

The depth of the centroid $= \frac{1}{2}(2c + b\cos\theta)$, and the whole pressure $= \frac{1}{2}w(2c + b\cos\theta)ab$.

Ex. 2. A vertical cylinder, radius r and height h, is filled with fluid.

The surface $= 2\pi rh$, the depth of the centroid $= \frac{1}{2}h$, and therefore the whole pressure $= w\pi rh^2$.

Ex. 3. A hollow cone, vertex downwards, is filled with water.

Let r be the radius, and h the height of the cone.

By cutting the cone down a generating line and unrolling it into a plane, its surface forms the sector of a circle, of which the slant side is the radius and the perimeter of the base is the arc.

---

* See Greaves's *Statics*, or Parkinson's *Mechanics*.

But the area of a sector $=\frac{1}{2}$ (arc) (radius) ;

$$\therefore \text{ the surface } = \pi r \sqrt{r^2 + h^2}.$$

Again, the surface of a cone is the ultimate form of the surface of a pyramid formed by triangles, having the vertex of the cone as their common vertex, and having for their bases the sides of a polygon inscribed in the circle, and since the centroid of each triangle is at a depth $\frac{1}{3}h$ below the surface of the fluid, it follows that $\frac{1}{3}h$ is the depth of the centroid of the surface.

Hence the whole pressure $= \frac{1}{3}w\pi rh\sqrt{r^2 + h^2}$.

Ex. 4. The cylinder in Ex. 2, closed at both ends, is just filled with liquid, and its axis is inclined at an angle θ to the vertical.

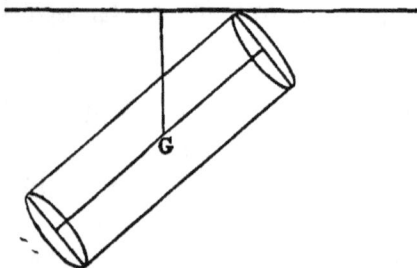

The surface of the fluid is a horizontal plane through the highest point of the cylinder, and the depth of G

$$= \frac{h}{2} \cos \theta + r \sin \theta.$$

Hence the whole pressure on the curved surface is

$$w\pi rh \ (h \cos \theta + 2r \sin \theta),$$

and the whole pressure including the plane ends is

$$w \ (\pi rh + \pi r^2)(h \cos \theta + 2r \sin \theta).$$

Ex. 5. A cubical vessel is filled with two liquids of given densities, the volume of each being the same, it is required to find the pressure on the base and on any side of the vessel.

Let a be a side of the vessel, w, w' the intrinsic weights of the upper and lower liquids, w' being taken greater than w.

The pressure on the base = the weight of the whole fluid $= w' \dfrac{a^3}{2} + w \dfrac{a^3}{2}$.

The pressure on the portion BC

$$= w \frac{a^2}{2} \cdot \frac{a}{4} = \frac{1}{8} wa^3.$$

To find the pressure on AC, replace the liquid DC by an equal weight of the lower liquid. This change will not affect the pressure at any point of CA.

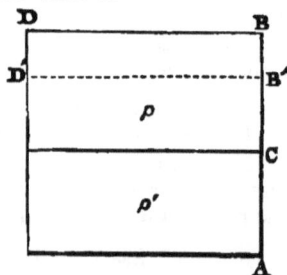

If $B'D'$ be its surface,

$$w'CB' = wCB = w\,\frac{a}{2},$$

and the depth of the centroid of AC below B'

$$= B'C + \frac{a}{4} = \frac{a}{4}\left(1 + \frac{2w}{w'}\right);$$

hence the pressure on $AC = w' \cdot \frac{a^2}{2} \cdot \frac{a}{4}\left(1 + \frac{2w}{w'}\right) = (w' + 2w)\frac{a^3}{8},$

and therefore the pressure on $AB = (3w + w')\frac{a^3}{8}$.

Centre of Pressure.

47. DEF. *The centre of pressure of a plane area is the point of action of the resultant fluid pressure upon the plane area.*

As a simple case, suppose a rectangle immersed in a liquid with one side in the surface.

Divide the area into a number of very small equal parts by equidistant horizontal lines.

The pressure on each part will act at its middle point and will be proportional to the depth below the surface, and we have to find the centre of a system of parallel forces acting perpendicularly to the plane at equidistant points of the line EF and proportional to the distance from E.

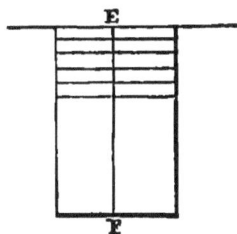

This is evidently the same as finding the centroid of a triangle of which E is the vertex and F the middle point of the base. The centre of pressure therefore divides EF in the ratio $2 : 1$.

It will be seen that this result is independent of the inclination of the plane of the rectangle to the vertical.

If a triangular area be immersed with its vertex in the surface and its base horizontal, and be divided by equidistant horizontal lines, the pressure on each strip will act at its middle point and be proportional to the square of the distance of that point from the vertex E.

Hence if F be the middle point of the base, the centre of pressure will be the same as the centroid of a solid cone, vertex E and axis EF, and therefore divides EF in the ratio $3 : 1$.

If a triangular area be immersed with its base in the surface, the pressure on a strip will be proportional to the product $EN . NF$, and consequently proportional to the square of the ordinate NP of a semi-circle described upon EF as diameter.

The centre of pressure will therefore be the middle point of EF.

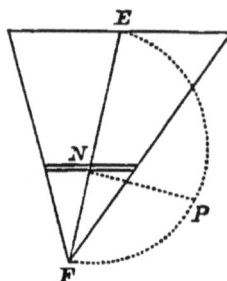

48. We may also give the following general method, applicable to the case of a plane area immersed in any position.

Through the boundary line of the plane area draw vertical lines to the surface and consider the equilibrium of the liquid so enclosed; the reaction of the plane resolved vertically, is equal to the weight of the liquid, which acts in the vertical line through its centre of gravity; and the point in which this line meets the plane is the centre of pressure.

Bearing in mind the fact that the pressure is proportional to the depth below the surface, it will be seen that the depth of the centre of gravity of the liquid thus enclosed is one half of the depth of the centre of pressure of the plane area.

49. If a plane be immersed vertically, and then turned, through any angle, round its line of intersection with the surface of the liquid, the pressures at all its points will be changed in the same ratio.

It follows therefore that such a rotation will not affect the position of the centre of pressure of any area upon the plane.

50. If a plane area is immersed vertically to a given depth, and if the position of its centre of pressure is known,

we can determine the position of the centre of pressure for any other given depth.

Let K be the position of the centre of pressure, when G, the centroid of the area, is at the depth h.

If the depth is increased to h', the increase of pressure on the area A acts at G and is equal to

$$wA \, (h' - h).$$

Take the point K' in GK such that

$$wA \, (h' - h) \cdot GK' = wAh \cdot KK',$$

or $\qquad\qquad h' \cdot GK' = h \cdot GK.;$

then K' is the new centre of pressure.

51. The student will now be able to appreciate more clearly the nature of fluid pressures, and to see that the action of a fluid does not depend upon its quantity, but upon the position and arrangement of its continuous portions. It must be carefully borne in mind that the surface of an inelastic fluid or liquid is always the horizontal plane drawn through the highest point or points of the fluid, and that the pressure depends only on the depth below that horizontal plane.

Thus in the construction of dock-gates, or canal-locks, it is not the expanse of sea outside which will affect the pressure, but the height of the surface; and, in considering the strength required in the construction, the greatest height of the surface due to tides must also be taken into account. Any violent action due to rapid tides or storms is of course a subject for separate consideration.

The same principle shews that in the construction of dikes, or the maintenance of river-banks, the strength must be proportional to the depth below the surface.

EXAMINATION UPON CHAPTER III.

1. To what extent is the pressure on the base of a vessel affected by pouring in more liquid?

2. Find the pressure at a depth of 100 feet in a lake, 1st, neglecting, 2nd, taking account of the atmospheric pressure.

3. Explain the statement that liquids maintain their level.

4. A reservoir of water is 200 feet above the level of the ground-floor of a house; find the pressure of the water in a pipe at a height of 30 feet above the ground-floor.

5. Three liquids that do not mix are contained in a vessel; prove that their common surfaces are horizontal, and find the pressure at any depth in the lowest liquid.

6. An equilateral triangular area is immersed in water with a side two feet in length in the surface; find the pressure upon it.

7. Distinguish between whole pressure and resultant pressure.

8. A hollow cone, vertex upwards, is just filled with liquid; find the whole pressure on its curved surface.

9. Prove that the depth of the centre of pressure of a plane area is greater than the depth of the centre of gravity of the area.

10. Find the centre of pressure of a rectangular area immersed, with plane vertical and two sides horizontal.

11. A rectangle has one side in the surface of a liquid; divide it by a horizontal line into two parts on which the pressures are equal.

12. Divide the same rectangle by horizontal lines into n parts on which the pressures are equal.

13. A triangle has its base horizontal and its vertex in the surface; divide it by a horizontal line into two parts on which the pressures are equal.

EXAMPLES.

1. Two equal vertical cylinders standing on a horizontal table are connected together by a pipe passing close to the table, and are partially filled with water. In contact with and above the water in one cylinder is a closely-fitting piston of given weight; find its position of equilibrium.

2. The upper surface of a vessel filled with water is a square whose side is 2 feet 6 inches, and a pipe communicating with the interior is filled with water to a height of 8 feet; find the weight which must be placed on the lid of the vessel to prevent the water from escaping.

3. A parallelogram is immersed in a liquid with one side in the surface; shew how to draw a line from one extremity of this side dividing the parallelogram into two parts on which the pressures are equal.

4. A fine tube ABC is bent so that the portions AB, BC are straight and perpendicular to each other; the tube is placed so that each branch is equally inclined to the vertical, and equal quantities of two liquids, the densities of which are in the ratio of 2 : 1, are poured into the respective branches; find the height above B of their common surface.

5. A smooth vertical cylinder one foot in height and one foot in diameter is filled with water, and closed by a heavy piston weighing 4 lbs.; find the whole pressure on its curved surface.

6. If a ball, weighing 1 lb. in water, be suspended in the water by a string fastened to the piston, and if the height of the piston above the base be still one foot, find the pressure at any depth and the whole pressure on the curved surface.

7. A cylindrical vessel standing on a table contains water, and a piece of lead of given size supported by a string is dipped into the water; how will the pressure on the base be affected, (1) when the vessel is full, (2) when it is not full? and in the second case, what is the amount of the change?

8. A hollow cylinder closed at both ends is just filled with water and held with its axis horizontal: if the whole pressure on its surface, including the plane ends, be three times the weight of the fluid, compare the height and diameter of the cylinder.

9. A triangle ABC is immersed vertically in a liquid with the angle C in the surface and the sides AC, BC equally inclined to the surface; shew that the vertical through C divides the triangle into two others, the fluid pressures upon which are as $b^3 + 3ab^2 : a^3 + 3a^2b$.

10. A vertical cylinder contains equal volumes of two liquids, the intrinsic weight of the lower liquid being three times that of the upper liquid; find the whole pressure on the curved surface, and prove that, if the fluids be mixed together so as to become a homogeneous mass, the whole pressure will be increased in the ratio of 4 to 3.

11. A triangle is immersed in a fluid with one of its sides in the surface; find the position of a point within the triangle, such that, if it be joined to the angular points, the triangle shall be divided into three others, the fluid pressures upon which are equal.

12. The side AB of a triangle ABC is in the surface of a fluid, and a point D is taken in AC, such that the pressures on the triangles BAD, BDC, are equal; find the ratio $AD : DC$.

13. The lighter of two fluids, whose specific gravities are as 2 : 3, rests on the heavier, to a depth of four inches. A square is immersed in a vertical position with one side in the upper surface; determine the side of the square in order that the pressures on the portions in the two fluids may be equal.

14. A vertical cylinder contains equal portions of three inelastic fluids, of intrinsic weights, w, $2w$, $3w$, respectively, the lighter fluid being uppermost, and the heavier fluid lowest; compare the whole pressures on the portions of the curved surface of the cylinder in contact with the several fluids.

15. A fine tube, which is bent into the form of a circle, contains given quantities of two different liquids; if the two together occupy half the tube, determine the position of equilibrium.

16. The inclinations of the axis of a submerged solid cylinder to the vertical in two different positions are complementary to each other; P is the difference between the pressures on the two ends in the one, and P' in the other position: prove that the weight of the displaced fluid is equal to

$$(P^2 + P'^2)^{\frac{1}{2}}.$$

17. A vertical cylinder contains a quantity of fluid, whose depth equals a diameter of the circular base. A sphere of four times the intrinsic weight of the fluid and of the same radius as the cylinder is placed upon the fluid and is supported by it: find the increase of pressure sustained by the curved surface of the cylinder, the sphere fitting it exactly.

18. Three fluids whose densities are in arithmetic progression, fill a semicircular tube whose bounding diameter is horizontal. Prove that the depth of one of the common surfaces is double that of the other.

19. Prove that, as a plane area is lowered vertically in a liquid, the centre of pressure approaches to, and ultimately coincides with, the centre of gravity.

20. A circular area is just immersed vertically in water; prove that, if the depth of its centre is doubled, the distance between its centre and the centre of pressure will be halved.

21. A square lamina is just immersed vertically in water, and is then lowered through a depth b; if a is the length of the edge of the square prove that the distance of the centre of pressure from the centre of the square will be

$$a^2/(6a + 12b).$$

22. A lamina in the shape of a quadrilateral $ABCD$ has the side CD in the surface, and the sides AD, BC vertical and of lengths a, β, respectively. Prove that the depth of the centre of pressure is

$$\frac{1}{2} \cdot \left(\frac{a^3 + a^2\beta + a\beta^2 + \beta^3}{a^2 + a\beta + \beta^2} \right).$$

23. A vessel contains two liquids whose densities are in the ratio of 1 to 14. A triangle is immersed vertically in the liquids so that its base is in the surface of the upper liquid. If the pressures on the portions in the two liquids be equal, prove that the areas of those portions are as 8 to 1.

24. The depth of the water on one side of a rectangular vertical floodgate is double that on the other. Supposing the gate to be fastened at the angular points, find the pressures at these points.

25. A vertical cylinder contains equal quantities of two liquids; compare their densities when the whole pressures of the two liquids on the curved surface of the cylinder are in the ratio 1 : 3.

26. Compare the whole pressures on the curved surface and plane base of a solid hemisphere, which is just immersed in water with its base horizontal and downwards.

27. Find the centre of pressure of a square just immersed in a liquid with one diagonal vertical.

28. Prove that whatever be the law of density of a liquid contained in a right circular cone with its axis vertical and vertex upwards the whole pressure is the same as if the fluid were mixed up so as to become of uniform density.

29. Prove that the depth of the centre of pressure of a trapezium immersed in water with the side a in the surface, and the parallel side b at a depth h below the surface is

$$\frac{a + 3b}{a + 2b} \cdot \frac{h}{2}.$$

30. A closed hollow cone is just filled with liquid, and is placed with its vertex upwards and axis vertical; divide its curved surface by a horizontal plane into two parts on which the whole pressures are equal.
Also do the same when the vertex is downwards.

31. If three liquids which do not mix, and whose densities are ρ_1, ρ_2, ρ_3, fill a circular tube in a vertical plane, and if a, β, γ are the angles which the radii to the common surfaces make with the vertical diameter measured in the same direction, prove that

$$\rho_1 (\cos\beta - \cos\gamma) + \rho_2 (\cos\gamma - \cos a) + \rho_3 (\cos a - \cos\beta) = 0.$$

If there are equal quantities of each fluid, and if in addition the weights on each side of the vertical diameter are equal, obtain an equation to determine a, which refers to the highest point of junction. Shew that it is satisfied by $a = 30^0$, and that therefore the densities are in arithmetic progression.

32. A solid triangular prism, the faces of which include angles
a, β, γ, is completely immersed in water with its edges horizontal; if
P, Q, R, be the pressures on the three faces, which are respectively
opposite to the angles a, β, γ, prove that

$$P \operatorname{cosec} a + Q \operatorname{cosec} \beta + R \operatorname{cosec} \gamma$$

is invariable so long as the depth of the centre of gravity of the prism
is unchanged.

33. A cubical vessel, standing on a horizontal plane, has one of its
vertical sides loose, which is capable of revolving about a hinge at the
bottom.   If a portion of fluid equal in volume to one-fourth of the
cube be poured into the vessel, the loose side will rest at an inclination
of 45^0 to the horizon: compare the weight of the side with the weight
of the fluid in the vessel.

34. A cubical box, filled with water, has a close fitting heavy lid
fixed by smooth hinges to one edge; compare the tangents of the
angles through which the box must be tilted about the several edges of
its base, in order that the water may just begin to escape.

35. A cylindrical tumbler, containing water, is filled up with wine;
after a time half the wine is floating on the top, half the water remains
pure at the bottom, and the middle of the tumbler is occupied by wine
and water completely mixed, the common surfaces being horizontal
planes; if the weight of the wine be two-thirds of that of the water, and
their densities be in the ratio of 11 : 12, prove that in this position the
whole pressure of the pure water on the curved surface of the tumbler
is equal to the whole pressure of the remainder of the liquid on the
tumbler.

36. A cone, with its axis inclined at an angle θ to the vertical,
contains some water; it is turned till its axis is vertical.   Shew that
the whole pressure is altered in the ratio $\cos \theta : 1$.

37. An oblique cylinder standing on a horizontal plane, the
generating lines making an angle a with the vertical, is filled to a
height h with a weight W of liquid.   Prove that the resultant pressure
on the curved surface of the cylinder is equivalent to a couple of
moment $\frac{1}{2} Wh \tan a$, tending to upset the cylinder.

CHAPTER IV.

RESULTANT VERTICAL AND HORIZONTAL PRESSURE ON ANY
SURFACE, RESULTANT PRESSURE ON THE SURFACE OF
AN IMMERSED SOLID, CONDITIONS OF EQUILIBRIUM OF
A FLOATING BODY, THE CAMEL, METHOD OF REMOVING
WOODEN PILES, STABILITY OF EQUILIBRIUM, META-
CENTRE, BODIES FLOATING IN AIR, THE BALLOON.

52. PROP. *To find the resultant vertical pressure of a liquid on any surface.*

Let PQ be a portion of surface in contact with a liquid at rest, and through the boundary line of PQ draw vertical lines to the surface AB, thus enclosing a mass of the liquid.

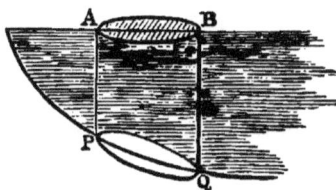

The pressure of the sur-
rounding liquid on this mass is entirely horizontal, and it is therefore clear that the weight of the mass is entirely supported by the reaction of the surface PQ.

Hence the vertical component of this reaction must be equal to the weight of the mass $ABQP$.

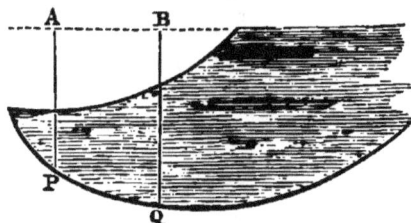

By the previous Chapter this is true whether the curve AB be really in the liquid, or only in the horizontal plane through the highest point of the liquid, as in the figure.

Hence it follows that *the resultant vertical pressure is the weight of the superincumbent liquid.*

The phrase superincumbent liquid must be interpreted, in the second figure, as denoting the mass of liquid which would occupy the space $PAQB$.

53.   There are other cases which it is requisite to consider.

Thus the liquid may press *upwards* on the surface.

In this case, let AB as before be the curve formed by vertical lines round PQ, and imagine the liquid within to be removed and the outside of PQ to be under the pressure of a fluid of which AB is the surface. It will be seen that the pressure at any point of PQ is the same as before in magnitude, but opposite in direction, and the resultant vertical pressure is therefore the same, only that it is now downwards, and by the previous article it is equal to the weight of $ABQP$.

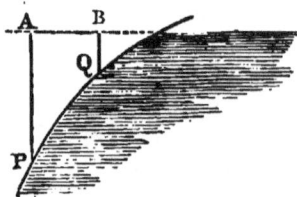

Hence the resultant vertical pressure upwards on PQ is as before equal to the weight of the liquid above it, that is, between PQ and the surface.

Or the pressure may be partly upwards and partly downwards, as on PEQ.

Draw QQ' vertical, and consider the pressures on QEQ' and $Q'P$ separately.

By the same reasoning the vertical pressure on QEQ' is downwards and equal to the weight of the liquid contained between the surface and the vertical plane QQ', and the difference between this and the upward vertical pressure on PQ' is the resultant vertical pressure downwards on the surface PQ.

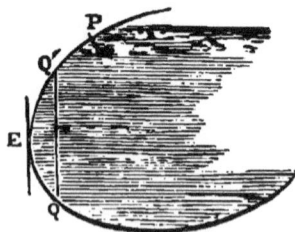

In all cases the line of action of the resultant vertical pressure is the vertical through the centre of gravity of the superincumbent liquid.

54. PROP. *To find the resultant horizontal pressure in a given direction of a liquid on any surface.*

Take a fixed vertical plane perpendicular to the given direction, and draw horizontal lines through the boundary of the surface PQ, meeting the vertical plane in the curve AB. The equilibrium of the liquid thus enclosed is maintained by its own weight, by the fluid pressures on its curved surface which are all parallel to the vertical plane, and by the fluid pressures on the surfaces AB and PQ.

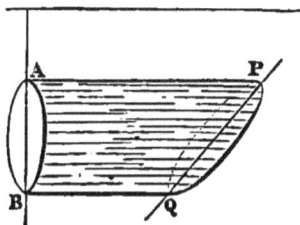

Hence the horizontal component of the reaction of PQ must be equal to the pressure on AB, which can be found from previous investigations, and the line of action will be the horizontal line through the centre of pressure of AB.

55. We are now in a position to determine the resultant pressure in direction and magnitude of a liquid on any surface; for we can obtain separately the vertical and horizontal pressures, and hence, by the principles of Statics, determine the magnitude and direction of the resultant.

Ex. 1. A vessel in the form of an open semi-cylinder with its ends vertical, is filled with water; it is required to find the resultant pressure on either of the portions into which it is divided by a vertical plane through the axis of the cylinder.

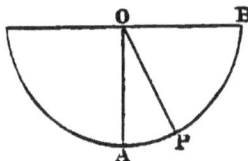

Let h be the length of the cylinder and a its radius, and let the figure be a vertical section through the middle point O of its length.

The resultant vertical pressure on AB

= the weight of the fluid OAB

$= wh\,\dfrac{\pi a^2}{4}$, if w is the intrinsic weight of the water.

The resultant horizontal pressure on AB = the pressure on the vertical section perpendicular to the plane of the paper, that is, on a rectangle of which the sides are a and h,

$$= wah\,\frac{a}{2} = \frac{1}{2}\,wa^2h.$$

Hence the angle θ, at which the direction of the resultant pressure is inclined to the horizon, is given by the equation

$$\tan \theta = \frac{\frac{1}{4} w\pi a^2 h}{\frac{1}{2} wa^2 h} = \frac{\pi}{2}.$$

Moreover, since the pressure at any point acts in a direction passing through the axis of the cylinder, the resultant pressure acts in a line through O, and, if $POB = \tan^{-1}\left(\frac{\pi}{2}\right)$, the point P is the centre of pressure of the curvilinear surface.

Ex. 2. *A closed hemispherical vessel is just filled with liquid, and is held with its plane base vertical.*

Consider the equilibrium of the liquid, and observe that the resultant horizontal and vertical pressures of the curved surface on the liquid are respectively

$$w\pi a^3 \text{ and } \frac{2}{3} w\pi a^3.$$

Hence, if θ is the inclination to the vertical of the line of action of the resultant pressure of the curved surface on the liquid,

$$\tan \theta = \frac{3}{2}.$$

We can hence obtain the position of the centre of pressure of the plane base.

For the lines of action of the resultant pressures of the curved surface, of the weight, and of the resultant pressure of the plane base must be concurrent, and therefore since the distance of the centroid of the liquid from the centre of the sphere is three-eighths of the radius, it follows that the depth of the centre of pressure of the plane base below its centre is one-fourth of the radius.

Ex. 3. A hollow cone filled with water is held with its vertex downwards; it is required to determine the resultant pressure on either of the portions into which it is divided by a vertical plane through its axis.

Let a be the radius of the base and $2a$ the vertical angle.

The volume $= \frac{1}{3} \pi a^3 \cot a.$

The resultant vertical pressure on the portion $AEVB$

$= \frac{1}{2}$ the weight of the fluid

$= \frac{1}{6} w\pi a^3 \cot a.$

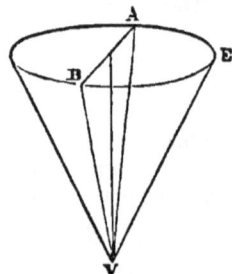

The resultant horizontal pressure

= the pressure on the triangle $A\,VB$

$$= w \,.\, a^2 \cot a \,.\, \frac{1}{3}\, a \cot a$$

$$= \frac{1}{3}\, wa^3 \cot^2 a\,;$$

therefore the resultant pressure

$$= \frac{w}{3}\, a^3 \cot a\, \sqrt{\frac{\pi^2}{4} + \cot^2 a},$$

and if θ be the angle at which its direction is inclined to the horizon,

$$\tan \theta = \frac{\frac{1}{6}\,\pi}{\frac{1}{3}\,\cot a} = \frac{\pi}{2}\, \tan a.$$

In general the determination of the line of action can only be effected by means of the Integral Calculus, but in the first example we were able to infer at once the position of the line of action, and in some cases it may be determined by special geometrical contrivances.

As an example, the position of the line of action in this last case will be obtained in Chapter XII. by the help of such a contrivance.

56. *To find the resultant pressure of a liquid on the surface of a solid either wholly or partially immersed.*

Imagine the solid removed, and the space it occupied in the liquid to be filled with the liquid. It is clear that the resultant pressure on this liquid is the same as on the original solid. The weight of this liquid is entirely supported by the pressure of the surrounding liquid, and therefore the resultant pressure is equal to the weight of the displaced liquid, and acts vertically upwards in a line passing through its centre of gravity.

This is sometimes expressed by saying that *a solid immersed in fluid loses as much of its weight as is equal to the weight of the fluid it displaces,* observing that the above reasoning is equally applicable to the case of a body immersed in elastic fluid.

57. *A solid of given volume V, having for a part of its surface a plane of given area A, is completely immersed in a liquid; having given the depth z, of the centroid of this area*

and its inclination, θ, to the vertical, it is required to find the resultant pressure on the remainder of the surface of the solid.

If the plane area is on the upper surface of the solid, as in the figure, the resultant horizontal pressure on the plane area is $wAz \cos \theta$, and the resultant vertical pressure, downwards, is $wAz \sin \theta$.

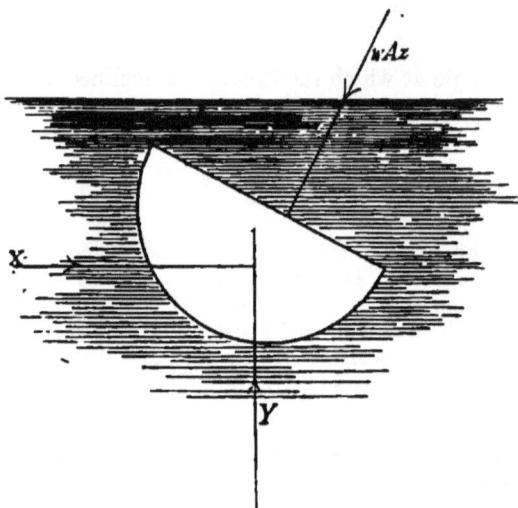

The resultant pressure on the whole surface is vertical and is equal to wV; ∴ if X and Y are the resultant horizontal and vertical pressures on the rest of the surface, Y being measured upwards,

$$X = wAz \cos \theta,$$

and

$$Y - wAz \sin \theta = wV.$$

If the plane area forms part of the lower boundary of the surface of the solid, so that the vertical pressure is upwards, the second equation will take the form

$$wAz \sin \theta - Y = wV,$$

Y being now measured downwards.

Hence the actual resultant pressure on the rest of the surface, which is $\{X^2 + Y^2\}^{\frac{1}{2}}$, is equal to

$$w \{A^2 z^2 \pm 2AzV \sin \theta + V^2\}^{\frac{1}{2}},$$

the upper sign belonging to the first case, and the lower sign to the second case.

58. *To find the conditions of equilibrium of a floating body.*

We have shewn, in article (56), that the resultant pressure of the liquid on the surface of the body is equal to the weight of the displaced liquid, acting vertically upwards.

It follows therefore that, the body being supported entirely by the liquid, the weight of the displaced liquid must be equal to the weight of the body, and the centres of gravity of both must be on the same vertical line.

These conditions also hold good when the body floats partly immersed in two or more liquids, and are, for such cases, established by precisely the same reasoning.

59. *If a homogeneous body float in a liquid, its volume will bear to the volume immersed the inverse ratio of the specific gravities of the solid and liquid.*

For if V, V' be the volumes, and s, s' the specific gravities,

$$62 . 5 \ Vs = \text{the weight of the body}$$
$$= \text{the weight of the displaced fluid}$$
$$= 62 . 5 \ V's',$$
$$\therefore \ V : V' = s' : s.$$

60. *To find the conditions of equilibrium of a solid floating in liquid and partly supported by a string.*

First, let the solid be homogeneous and wholly immersed; then the centres of gravity of the solid and of the liquid displaced will be the same, and the direction of the string must be the vertical through the centre of gravity. Also

the tension = the weight of the body − the weight lost

$$= V(s - s') \times 62\cdot5 \text{ lbs. weight,}$$

if s, s' be the specific gravities of the solid and fluid.

Secondly, let the solid be homogeneous and partly immersed.

Let V' be the part immersed, H its centre of gravity, and G the centre of gravity of the body.

Draw vertical lines through H and G meeting the surface in C and A, and let the direction of the string meet the surface in B.

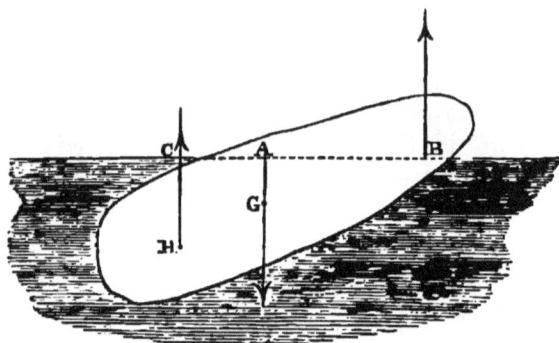

Then, if T be the tension, we have three equilibrating forces acting in the vertical lines through A, B, and C;

$$\therefore T = 62\cdot5 \left(Vs - V's' \right) \text{ lbs. weight.}$$

and
$$Vs \cdot AB = V's' \cdot CB.$$

The second equation is the condition of equilibrium, and the first gives the requisite tension.

The case in which a heterogeneous body is partly supported by a string may be left for the consideration of the student.

61. *The Camel.* This is an apparatus for carrying a ship over the bar of a river. It consists of four, or a greater number, of watertight chests, which are filled with water, placed in pairs on opposite sides of the ship, and attached to the ship, or attached to each other by chains passing under the keel. If the water be then pumped out, the vessel will be lifted, and may be towed over the bar into deep water. The lifting power of the camel is the weight of water displaced by the chests, diminished by the weight of the whole apparatus.

62. *Removing wooden Piles.* It is sometimes necessary to remove entirely piles which have been driven down in

deep water; for instance, the piles employed to keep out water during the construction of a dock. After the water has been allowed to flow within the piles, they are sawn off to a convenient depth, and a barge is floated over them and filled with water. The barge is then attached by chains to a pile, and the water pumped out; as the pumping proceeds the barge is lifted, and the pile is forcibly drawn out. If the operation take place in the sea, a great advantage is gained by fastening the barge to the pile at low tide. The rise of the tide will sometimes draw out the pile, but, if necessary, additional force must be gained by pumping water out of the barge.

63. We now proceed to exemplify the preceding propositions by their application to some particular cases.

Ex. 1. A man, whose weight is the weight of 150 lbs., and specific gravity 1·1, just floats in water, the specific gravity of which is 1, by the help of a quantity of cork. The specific gravity of cork being ·24, find its volume in cubic feet.

Let V be the volume of the cork, and V' of the man, in cubic feet.

Then $62·5 \{V(·24) + V'(1·1)\}$ = weight of water displaced

$$= 62·5 (V + V'),$$

or $$V(·76) = V'(·1).$$

Also $$62·5\, V'(1·1) = \text{weight of man} = 150,$$

$$\therefore \quad V' = \frac{24}{11},$$

and $$V = \frac{·1}{·76} \times \frac{24}{11} = \frac{60}{209} \text{ths of a cubic foot.}$$

Ex. 2. A cylindrical piece of wood floats in water with its axis vertical; find how much it will be depressed by placing a given weight on the top of it.

If w be the weight placed on the top, it will be depressed through such a space that the additional amount of displaced fluid has its weight equal to w.

Now, if W be the weight of the cylinder, it is also the weight of the fluid displaced by the cylinder, and therefore, if h be the depth of the base of the cylinder originally, and x the depression,

$$w : W :: x : h;$$

$$\therefore \quad x = \frac{w}{W} h.$$

If this value of x exceed the height of the cylinder originally above the surface, it will be entirely immersed, and the possibility of equilibrium will then depend on the density of the weight.

Ex. 3. An isosceles triangular lamina floats in liquid with its base horizontal: it is required to find the position of equilibrium when the base is above the surface.

Take ρ' and ρ as the densities of the lamina and of the liquid, h as the height of the triangle, and x the depth to which it is immersed.

Then, since the intrinsic weights of two bodies at the same place are proportional to their densities, it follows that ρ' (volume of lamina) $= \rho$ (volume of fluid displaced); and therefore, similar triangles being proportional to the squares of homologous sides, we have

$$\rho' h^2 = \rho x^2, \quad \text{and} \quad x = h \sqrt{\frac{\rho'}{\rho}}.$$

The second condition is obviously satisfied in this and the preceding example.

Ex. 4. Can an isosceles triangular lamina float with its base vertical in a liquid of twice its density?

The first condition requires that half the triangle should be immersed, and therefore its vertex A is in the surface.

Also, if G, H be the centres of gravity,

$$AG = \frac{2}{3} AE, \quad \text{and} \quad AH = \frac{2}{3} AF,$$

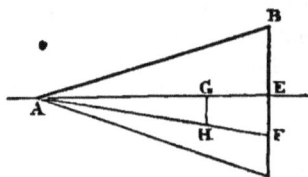

F being the middle point of EC;

$$\therefore \quad AG : AH :: AE : AF.$$

Hence GH is parallel to EF, is therefore vertical, and both conditions are satisfied.

Ex. 5. A cylinder floats with its axis vertical, partly immersed in two liquids, the densities of the upper and lower liquids being respectively ρ and 2ρ, and the density of the cylinder $\frac{3\rho}{4}$; find the position of equilibrium of the cylinder, its length being twice the depth of the upper fluid.

Let x be the length immersed in the lower fluid, k the area of either end, and $2h$ the whole length.

Then, intrinsic weights being proportional to densities,

$$\frac{3\rho}{4} k \cdot 2h = \rho k h + 2\rho k x; \quad \text{and} \quad \therefore \quad x = \frac{1}{4} h.$$

If the cylinder were just immersed, its density ρ' would be such that

$$2\rho' h = \rho h + 2\rho h;$$

or
$$\rho' = \frac{3\rho}{2},$$

and x would then be equal to h.

Ex. 6. A cubical box, the volume of which is one cubic foot, is three-fourths filled with water, and a leaden ball, the volume of which is 72 cubic inches, is lowered into the water by a string; it is required to find the increase of pressure on the base and on a side of the box.

The complete immersion of the lead will raise the surface $\frac{1}{2}$ an inch, since 144 square inches is the area of the surface.

The pressure on the base is therefore increased by the weight of 72 cubic inches of water, i.e. by the weight of $\frac{72}{1728}$ 1000 oz., or $41\frac{2}{3}$ oz.

The area of a side originally in contact with the fluid was $\frac{3}{4}$ of a square foot,

and the pressure was $1000 \times \frac{3}{4} \times \frac{3}{8}$ oz. wt., or $281\frac{1}{4}$ oz. wt.,

$\frac{3}{8}$ths of a foot being the depth of the centre of gravity.

The new area is $\frac{3}{4} + \frac{1}{24}$, or $\frac{19}{24}$ of a square foot;

$\therefore$ the new pressure $= 1000 \times \frac{19}{24} \times \frac{19}{48}$ oz. wt.

$$= 313\frac{53}{144} \text{ oz. wt.}$$

The increase is therefore a little more than the weight of 32 oz.

Ex. 7. *A solid hemisphere is moveable about the centre of its plane base which is fixed in the surface of a liquid; if the density of the liquid be twice that of the solid, any position of the solid will be one of rest.*

Hold the solid in the position ADB, DCE being the surface of the liquid, and continue the sphere to the surface E of the liquid.

Also make the angle DCF equal to the angle ECB, the figure being a vertical section through the centre C of the hemisphere perpendicular to its plane base.

Consider first the equilibrium of the mass BCE of liquid; this is maintained by the normal pressures on the surface BE, the reaction of the plane CB, and the weight of the liquid.

Hence it follows that the moment of the reaction of CB about the horizontal straight line through C perpendicular to the plane of the figure is equal to the moment, about the same line, of the weight of the liquid BCE. Next, considering the hemisphere, the wedge or

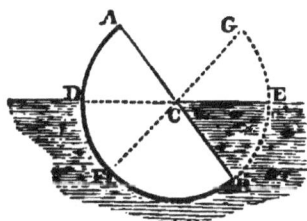

lune FCB would be of itself in equilibrium, and therefore the moment about C of the weight of ACF is equal to the moment of the fluid pressure upon CB.

Now we have shewn that the moment of this fluid pressure about C is equal to the moment about C of the weight of the mass CBE, and it is easily seen that the horizontal distance from C of the centre of gravity of BCE is equal to that of the centre of gravity of ACF.

Hence since the weight of ACF is equal to the weight of BCE, it follows that the forces on the hemisphere equilibrate, and therefore, releasing the hemisphere, it remains at rest.

The result of this problem has been practically employed in the construction of an oil-lamp, called Cecil's Lamp, such that the surface of the oil supplying the wick is always the same. DEB is a hemispherical vessel containing oil, and ADB a hemisphere, the specific gravity of which is half that of the oil; as the oil is consumed, ADB turns round C, and CE is always the surface of the oil.

Stability of Equilibrium.

64. Imagine a floating body to be slightly displaced from its position of equilibrium by turning it round so that the line joining its centre of gravity with that of the fluid displaced shall be inclined to the vertical. If the body on being released return to its original position its equilibrium is stable; if, on the other hand, it fall away from that position its original position is said to be one of unstable equilibrium.

Metacentre. In the figure let G, H be the centres of

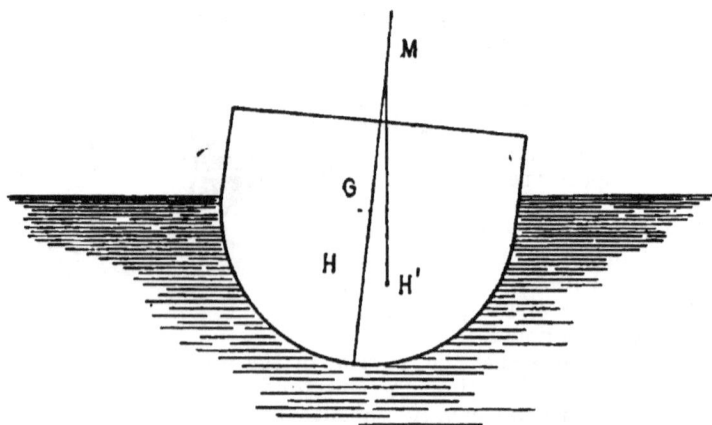

gravity of the body and of the fluid originally displaced, H' the centre of gravity of the fluid displaced in the new position, and M the point where a vertical through H' meets HG.

The resistance of the fluid acts vertically upwards in the line $H'M$, and it is therefore evident that, if M be above G, the action of the fluid will tend to restore the body to its original position; but, if M be below G, to turn the body farther from its original position.

The position of the point M will in general depend on the extent of displacement. If the displacement be very small, that is, if the angle between GH and the vertical be very small, the point M is called the *metacentre*, and the question of stability is now reduced to the determination of this point.

One of the most important problems in naval architecture is to secure the ascendancy under all circumstances of the metacentre over the centre of gravity.

This is effected by a proper form of the midship sections, so as to raise the metacentre as much as possible, and by ballasting so as to lower the centre of gravity, and the greater the distance between the points G and M, the greater is the steadiness of the vessel.

Moreover, the naval architect must have in view the probability of large displacements, due to the rolling of the vessel, and not merely the small movement which is considered in the determination of the metacentre.

65. In particular cases the metacentre can be sometimes found by elementary methods, but its general determination involves the application of the Integral Calculus.

In one case however its position is obvious. Let the lower portion of the solid be spherical in form; then as long as the portion immersed is spherical, the pressure of the water at every point acts in the direction of the centre of the sphere, and therefore the resultant pressure must act in the vertical line through the centre (E) of the sphere.

Now in the original position the centre of gravity of the fluid displaced is evidently in the vertical through E, and therefore the centre of gravity of the body is in the vertical through E.

Hence the point E is the metacentre.

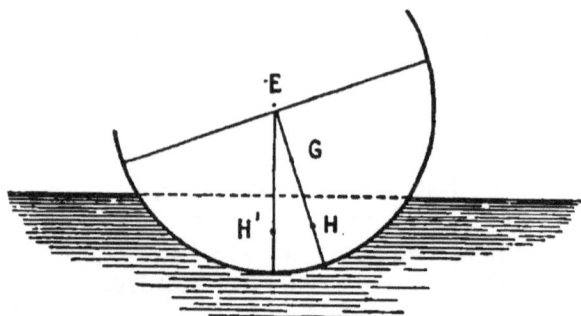

Thus if any portion whatever be cut from a solid sphere it will float in stable equilibrium with its curved surface partly immersed.

66. *Bodies floating in air.*

The fact that air is heavy enables us to extend to bodies, floating either wholly or partly in air, the laws of equilibrium which have been established for bodies floating in liquids.

Taking one case, if a body, lighter than water, float on its surface, it displaces a certain quantity of water and also a certain quantity of air; if we remove the body and suppose its place filled by air and water, it is clear that the weight of the displaced air and water is supported by the resultant vertical pressures of the air and water around it.

Hence the weight of the body must be equal to the weight of the air and water it displaces, and the centre of gravity of the air and water displaced must be in the same vertical line with the centre of gravity of the body.

In a similar manner, if a body float in air alone, its weight must be equal to the weight of the air it displaces.

67. In order to illustrate the case of a body floating in air and water, imagine that a piece of cork or some very light substance is floating in a basin of water, that the whole

is placed under the receiver of an air pump, and that the receiver is, as far as is practically possible, exhausted of air.

The effect on the position of equilibrium of the cork is determined by the fact that the weight of the cork is equal to the sum of the weights of the air and water displaced, so that, if the air be removed, more water must be displaced; the cork will therefore sink in the water.

This may also be shewn in the following manner.

In removing the air, we remove the downward pressure of the air on the cork, and also the downward pressure of the air on the surface of the water. This latter pressure is transmitted through the water to the lower surface of the cork, so that the forces on the cork are its weight, the downward pressure of the air on its upper surface, the pressure due to the water above, and the pressure of the air transmitted through the water.

Hence, since the atmospheric pressure on the surface of the water is greater than the atmospheric pressure at any point above the surface of the water, (see Art. 76), it follows that in removing the air from the receiver we remove the downward and upward pressures on the cork, of which the latter pressure is the greatest.

The cork will therefore sink in the water.

68. *The Balloon.* The ascent of a balloon depends on the principle of the previous article. A balloon is a very large envelope, made of silk, or some strong and light substance, and filled with a gas of less density than the air, usually coal gas. A car is attached in which the aeronauts are seated, and the weight of air displaced being greater than the whole weight of the balloon and car, the balloon ascends, and will continue ascending until the air around is not of sufficient density to support its weight. In order to descend, a valve is opened and a portion of the gas allowed to escape.

The ascensional force on a balloon is the weight of the air it displaces diminished by the weight of the balloon itself.

69. If a body float in a liquid, the centre of gravity of the liquid displaced is called the *Centre of Buoyancy*.

If the body be moved about, so that the volume of liquid displaced remains unchanged, the locus of the centre of gravity of the displaced liquid is called the *Surface of Buoyancy*.

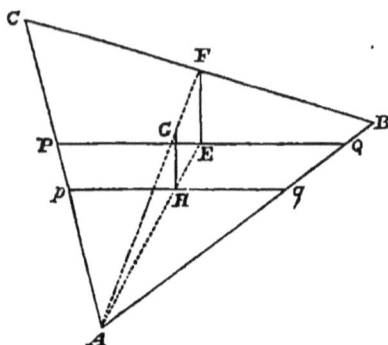

Taking a simple case, suppose a triangular lamina immersed with its plane vertical, and vertex beneath the surface, and let the area APQ be constant. Through H the centre of gravity of the area APQ, draw pHq parallel to PQ; then the area Apq is constant, and therefore pq always touches, at its middle point H, an hyperbola of which AB and AC are the asymptotes. This hyperbola is the curve of buoyancy. Now in the position of equilibrium, GH is vertical, and is consequently perpendicular to pq.

The position of equilibrium is therefore determined by drawing normals from G to the curve of buoyancy.

The problem then comes to the same thing as the determination of the positions of equilibrium of a heavy body, bounded by the surface of buoyancy, on a horizontal plane.

70. It is a general theorem that *the positions of equilibrium of a floating body are determined by drawing normals from the centre of gravity of the body to the surface of buoyancy.*

We give a proof for the case of a lamina with its plane vertical, or, which is the same thing, of a prismatic or cylindrical body with its flat ends vertical.

Let PQ, pq cut off equal areas, so that the triangles PCp, QCq are equal.

Then, if H be the centre of gravity of PAQ, E and F of PCp and QCq, take the point K in FH produced such that

$$KH : KF :: QCq : QAP;$$

and in KE the point H' such that

$$KH' : KE :: PCp : pAq;$$

then H' is the centre of gravity of pAq.

Hence, since $\quad KH' : KE :: KH : KF,$

it follows that HH' is parallel to EF, and therefore, ultimately, when the displacement is very small, HH' is parallel to PQ, or, in other words, the tangent to the curve of buoyancy at H is parallel to PQ.

Now, in the position of equilibrium, GH is vertical, and is therefore normal, at the point H, to the curve of buoyancy.

The *metacentre* having been defined as the point of intersection of the vertical through H' with the line HG, it follows that *the metacentre is the centre of curvature, at the point H, of the curve of buoyancy.*

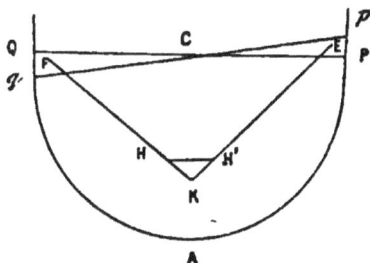

EXAMINATION UPON CHAPTER IV.

1. Shew how to find the resultant vertical pressure of a liquid on a surface; 1st, when it acts upwards, 2nd, when it acts downwards.

2. Apply the preceding to find the resultant pressure on a solid completely immersed.

3. A solid cone of metal, completely immersed in liquid, is supported by a string; find the tension of the string.

4. State the conditions of equilibrium of a floating body.

5. A wooden plank floats in water, and a weight is placed at one end of the plank; find the weight which, placed at a given distance from the other end, will keep the plank in a horizontal position.

6.   Describe a method of removing piles in deep water.

7.   A cylinder floats vertically in a fluid with 8 feet of its length above the fluid; find the whole length of the cylinder, the specific gravity of the fluid being three times that of the cylinder.

8.   A body floats in one fluid with ¾ths of its volume immersed, and in another with ⅘ths immersed; compare the specific gravities of the two fluids.

9.   A cylinder of wood 3 feet in length floats with its axis vertical in a fluid of twice its specific gravity; compare the forces required to raise it 6 inches and to depress it 6 inches.

10.   Three equal rods are jointed together so as to form an equilateral triangle, and the system floats in a liquid of twice the density of the rods, with one rod horizontal and above the surface; find the position of equilibrium.

11.   Explain what is meant by stability of equilibrium, and define the metacentre.

12.   A small iron nail is driven into a wooden sphere, and the weight of the sphere is then half that of an equal volume of water; find its positions of equilibrium in water, and examine the stability of the equilibrium.

13.   A sphere of ice floats in water, and gradually melts; does its centre rise or sink?

14.   A block of wood, the volume of which is 4 cubic feet, floats half immersed in water; find the volume of a piece of metal, the specific gravity of which is 7 times that of the wood, which, when attached to the lower portion of the wood, will just cause it to sink.

15.   A cylindrical block of wood is placed with its axis vertical in a cylindrical vessel whose base is plane, and water is then poured in to twice the height of the cylinder; find the pressure of the wood on the base of the vessel.

16.   Two cylindrical vessels, containing different fluids, and standing near each other on a horizontal plane, are connected by a fine tube, which is close to the horizontal plane; when the communication is opened between them, determine which of the fluids will flow from its own vessel into the other, and find the condition that the equilibrium may not be disturbed.

17.   Two bodies of given size and given specific gravities are connected by a string passing over a pulley, and rest completely immersed in water; find the condition of equilibrium.

NOTE ON CHAPTER IV.

The Principle of Archimedes. The enunciation and proof of the proposition of Article (58) are due to Archimedes, and it is a remarkable fact in the history of science, that no further progress was made in Hydrostatics for 1800 years, and until the time of Stevinus, Galileo, and Torricelli, the clear idea of fluid action thus expounded by Archimedes remained barren of results.

An anecdote is told of Archimedes, which practically illustrates the accuracy of his conceptions. Hiero, king of Syracuse, had a certain quantity of gold made into a crown, and suspecting that the workman had abstracted some of the gold and used a portion of alloy of the same weight in its place, applied to Archimedes to solve the difficulty. Archimedes, while reflecting over this problem in his bath, observed the water running over the sides of the bath, and it occurred to him that he was displacing a quantity of water equal to his own bulk, and therefore that a quantity of pure gold equal in weight to the crown would displace less water than the crown, the volume of any weight of alloy being greater than. that of an equal weight of gold. It is related that he immediately ran out into the streets, crying out εὕρηκα! εὕρηκα!

The two books of Archimedes which have come down to us, "*De iis quæ in humido vehuntur*" were first found in an old Latin MS. by Nicholas Tartaglia, and edited by him in 1537. In the first of these books it is shewn that the surface of water at rest must be a sphere of which the centre is at the earth's centre, and various problems are then solved relating to the equilibrium of portions of spherical bodies. The second book contains the proposition of Art. (58), and the solutions of a number of problems on the equilibrium of paraboloids, some of which involve complicated geometrical constructions.

The authenticity of these books is confirmed by the fact that they are referred to by Strabo, who not only mentions their title, but also quotes the second proposition of the first book.

Stevinus and Galileo. The Treatise of Stevinus on Statics and Hydrostatics, about 1585, follows that of Archimedes in the order of thought. In this he shewed how to determine the pressure of a liquid on the base and sides of a vessel containing it.

Galileo, in his Treatise on Floating Bodies, published in 1612, states the Hydrostatic paradox, and explains why the floating of bodies does not depend on their form.

EXAMPLES.

1. A uniform solid floats freely in a fluid of specific gravity twice as great as its own; prove that it will also float in equilibrium, if its position be inverted.

2. A block of ice, the volume of which is a cubic yard, is observed to float with $\frac{2}{5}$ths of its volume above the surface, and a small piece of granite is seen embedded in the ice; find the size of the stone, the specific gravities of ice and granite being respectively ·918 and 2·65.

3. An isosceles triangular lamina floats with its base horizontal and beneath the surface of a liquid of twice its density; find the position of equilibrium.

4. A solid cone floats with its axis vertical in a liquid the density of which is twice that of the cone; compare the portions of the axis immersed, 1st, when the vertex is upwards, 2nd, when it is downwards.

5. If w_1, w_2, w_3 be the apparent weight of a body in three liquids, the specific gravities of which are s_1, s_2, s_3, prove that

$$w_1(s_2 - s_3) + w_2(s_3 - s_1) + w_3(s_1 - s_2) = 0.$$

6. An equilateral triangular lamina suspended freely from A, rests with the side AB vertical, and the side AC bisected by the surface of a heavy fluid; prove that the density of the lamina is to that of the fluid as 15 to 16.

7. A vertical cylinder of density $\frac{7\rho}{4}$ floats in two liquids, the density of the upper liquid being ρ, and of the lower 2ρ; if the length of the cylinder be twice the depth of the upper liquid, find its position of rest.

8. A wooden rod is tipped with lead at one end; find the density of a liquid in which it will float at any inclination to the vertical; the weight of the lead being half that of the rod, and its size being neglected.

9. The weight of the unimmersed portion of a body floating in water being given, find the specific gravity of the body, in order that its volume may be the least possible. .

10. A cylindrical glass cup weighs 8 oz., its external radius is $1\frac{1}{2}$ inches, and its height $4\frac{1}{2}$ inches; if it be allowed to float in water with its axis vertical, find what additional weight must be placed in it, in order that it may sink.

11. A vessel in the form of half the above cylinder with both its ends closed, floats in water, with its ends vertical; find the additional weight which being placed in the centre of the vessel will cause it to be totally immersed.

12. A uniform rod, whose weight is W, floats in water in a position inclined to the vertical with a particle, of weight W', attached to its lower end; shew that, if the density of the water be four times that of the rod, half the length of the rod will be immersed.

13. A uniform rod floats partly immersed in water, and supported at one end by a string; prove that, if the length immersed remain unaltered, the tension of the string is independent of the inclination of the rod to the vertical.

14. A spherical shell, the internal and external radii of which are given, floats half immersed in water; find its density compared with the density of water.

15. A heavy hollow right cone, closed by a base without weight, is completely immersed in a fluid, find the force that will sustain it with its axis horizontal.

16. Find the position of equilibrium of a solid cone, floating with its axis vertical and vertex upwards, in a fluid of which the density bears to the density of the cone the ratio 27 : 19.

17. A rectangular lamina $ABCD$ has a weight attached to the point B, and floats in water with its plane vertical and the diagonal AC in the surface; prove that the specific gravity of the fluid is three times that of the lamina.

18. A solid paraboloid floats in a liquid with its axis vertical and vertex downwards; having given the densities of the paraboloid and the liquid, find the depth to which the vertex is submerged.

19. A ship sailing from the sea into a river sinks two inches, but after discharging 40 tons of her cargo, rises an inch and a half; determine the weight of the ship and cargo together, the specific gravity of sea-water being 1·025, and the horizontal section of the ship for two inches above the sea being invariable.

20. A cylindrical vessel of radius r and height h is three-fourths filled with water; find the largest cylinder of radius r' and specific gravity ·5 which can be placed in the water without causing it to run over, the axes of the cylinders being vertical and r' less than r.

21. A hollow cylinder is just filled with water, and closed, and is then held with its axis horizontal; find the direction and magnitude of the resultant pressure on the lower half of the curved surface. Also, if the cylinder be held with its axis vertical, find the direction and magnitude of the resultant pressure on the same surface.

22. A solid cylinder, one end of which is rounded off in the form of a hemisphere, floats with the spherical surface partly immersed : find the greatest height of the cylinder which is consistent with stability of equilibrium.

23. A body floating on an inelastic fluid is observed to have volumes P_1, P_2, P_3 respectively above the surface at times when the density of the surrounding air is ρ_1, ρ_2, ρ_3; shew that

$$\frac{\rho_2 - \rho_3}{P_1} + \frac{\rho_3 - \rho_1}{P_2} + \frac{\rho_1 - \rho_2}{P_3} = 0.$$

24. A hollow cubical box, the length of an edge of which is one inch, and the thickness one-eighteenth of an inch, will just float in water, when a piece of cork, of which the volume is 4·34 cubic inches, and the specific gravity ·5, is attached to the bottom of it. Find the specific gravity of tin.

25. A steamer, loading 30 tons to the inch in the neighbourhood of the water-line in fresh water, is found after a 10 days' voyage, burning 60 tons of coal a day, to have risen 2 feet in sea water at the end of the voyage; prove that the original displacement of the steamer was 5720 tons, taking a cubic foot of fresh water as 62·5 pounds and of sea water as 64 pounds.

26. A frustum of a right circular cone, cut off by a plane bisecting the axis perpendicularly, floats with its smaller end in a fluid and its axis just half immersed; compare the densities of the cone and fluid.

27. A solid cone and a solid hemisphere, which have their bases equal, are united together, base to base, and the solid thus formed floats in water with its spherical surface partly immersed; find the height of the cone in order that the equilibrium may be neutral.

28. Three uniform rods, joined so as to form three sides of a square, have one of their free extremities attached to a hinge in the surface of a fluid, and rest in a vertical plane with half the opposite side out of the fluid; shew that the specific gravity of the rods is to that of the fluid as 31 to 40.

29. A triangle ABC floats in a fluid with its plane vertical, the angle B being in the surface of the fluid and the angle A not immersed. Shew that the density of fluid : density of the triangle :: $\sin B$: $\sin A \cos C$.

30. A solid cone floats with its axis vertical and vertex downwards in an inelastic fluid; prove that, whatever be the density of the fluid, supposing it greater than that of the solid, the whole pressure on its curved surface is the same.

31. Two fluids are in equilibrium, one upon the other, the lower fluid having the greater specific gravity, and a solid cylinder, the height of which is equal to the depth of the upper fluid, is immersed with its axis vertical: the specific gravity of the cylinder being greater than that of the upper fluid, find the position of equilibrium.
What would be the effect of an increase in the density of the upper fluid? Will the equilibrium be stable or unstable for a vertical displacement?

32. Two equal uniform rods AB, BC are freely jointed at B, and are capable of motion about A, which is fixed at a given depth below the surface of a uniform heavy fluid. Find the position in which both rods will rest partly immersed, and shew that, in order that such a position may be possible, the ratio of the density of the rods to the density of the fluid must be less than $\dfrac{5}{9}$.

33. An equilateral triangle, ABC, of weight W and specific gravity σ, is moveable about a hinge at A, and is in equilibrium when the angle C is immersed in water and the side AB is horizontal. It is then

turned about A in its own plane until the whole of the side BC is in the water and horizontal; prove that the pressure on the hinge in this position

$$= 2\,\frac{1-\sqrt{\sigma}}{\sqrt{\sigma}}\,W.$$

34. If a floating body be wholly immersed, but different parts of it in any number of different liquids, shew that the specific gravities of the solids and liquids may be supposed to be so altered as to make one of the liquids a vacuum and another water without disturbing the equilibrium.

A piece of iron (S.G. 7·8) floats partly in two liquids (S.G. ·8 and 13·6): find the specific gravity of a solid which would float similarly in a vacuum and water. Hence find the ratio of the parts immersed.

35. Prove that a homogeneous solid, in the form of a right circular cone, can float in a liquid of twice its own density with its axis horizontal, and find, in that case, the whole pressure on the surface immersed.

36. A solid cone is just immersed with a generating line in the surface; if θ be the inclination to the vertical of the resultant pressure on the curved surface, prove that

$$(1 - 3\sin^2 a)\,.\,\tan\,\theta = 3\sin a\,.\,\cos a,$$

$2a$ being the vertical angle of the cone.

37. A hollow sphere is just filled with liquid; find the line of action and magnitude of the resultant pressure on either of the portions into which it is divided by a vertical plane through its centre.

38. A sphere is divided by a vertical plane into two halves which are hinged together at the lowest point, and it is just filled with homogeneous liquid; find the tension of a string which ties together the highest points of the two halves.

39. A right circular cone of height h and vertical angle $2a$, made of uniform material, floats in water with axis vertical and vertex downwards and a length h' of axis immersed. The cone is bisected by a vertical plane through the axis, and the two parts are hinged together at the vertex. Prove that the two halves will remain in contact if $h' > h\sin^2 a$.

40. A solid hemisphere is completely immersed with the centre of its base at a given depth; if W be the weight of fluid it displaces, P the resultant vertical pressure, and Q the resultant horizontal pressure, on its curved surface, prove that for all positions of the solid

$$(W - P)^2 + Q^2 \text{ is constant.}$$

41. A hollow cone, filled with water and closed, is held with its axis horizontal; find the resultant vertical pressure on the upper half of its curved surface.

42.   A solid cylinder which is completely immersed in water has its centre of gravity at a given depth below the surface, and its axis inclined at a given angle to the vertical ; determine the resultant horizontal and resultant vertical pressures upon its curved surface, and the direction and magnitude of the resultant pressure on the curved surface.

43.   The vertical angle of a solid cone is 60°; prove that it can float in a liquid with its vertex above the surface and its base touching the surface, if the densities of the cone and the liquid are in the ratio of $2\sqrt{2} - 1 : 2\sqrt{2}$.

44.   A thin hollow cone closed by an equally thin base will remain wherever it is placed entirely within a liquid ; prove that its vertical angle is $2 \operatorname{cosec}^{-1} 3$.

45.   The base of a vessel containing water is a horizontal plane, and a sphere of less density than water is kept totally immersed by a string fastened to the centre of a circular disc, which lies in contact with the base.   Find the greatest sphere of given density, and also the sphere of given size and least density, which will not raise the disc.

46.   In H.M.S. Achilles, a ship of 9000 tons displacement, it was found that moving 20 tons from one side of the deck to the other, a distance of 42 feet, caused the bob of a pendulum 20 feet long to move through 10 inches.   Prove that the metacentric height was $2\cdot24$ feet.

CHAPTER V.

ELASTICITY OF AIR, EFFECT OF HEAT, THERMOMETERS, TORRICELLI'S EXPERIMENT, WEIGHT OF AIR, THE BAROMETER AND ITS GRADUATION, THE RELATIONS BETWEEN PRESSURE, DENSITY, AND TEMPERATURE, DETERMINATION OF HEIGHTS BY THE BAROMETER, THE SIPHON, GRADUATION OF A THERMOMETER, THE DIFFERENTIAL THERMOMETER.

71. THE pressure of an elastic fluid is measured exactly in the same way as the pressure of a liquid, and it has been mentioned before that the properties of equality of pressure in all directions and of transmission of pressure are equally true of liquids and gases.

There is however this difference between a gas and a liquid, that the pressure of the latter is entirely due to its weight, or to the application of some external pressure, while the pressure of a gas, although modified by the action of gravity, depends in chief upon its volume and temperature.

The action of a common syringe will serve to illustrate the elasticity of atmospheric air. If the syringe be drawn out and its open end then closed, a considerable effort will be required to force in the piston to more than a small fraction of the length of its range, and if the syringe be air-tight, and strong enough, it will require the application of very great power to force down the piston through nearly the whole of its range. Moreover, this experiment with a syringe shews that the pressure increases with the compression, the air within the syringe acting as an elastic cushion. If the piston after being forced in be let go, it will be driven back, the air within expanding to its original volume.

Another simple illustration may be obtained by immersing carefully in water an inverted glass cylinder. Holding the cylinder vertical, fig. Art. 97, Ex. (2), it may be pressed down in the water without much loss of air, and it will be seen that the surface of the water within the vessel is below the surface of the water outside. It is evident that the pressure of the air within is equal to the pressure of the water at its surface within the cylinder, which, as we have shewn before, is equal to the pressure at the outside surface, increased by the pressure due to the depth of the inner surface; hence the air within, which has a diminished volume, has an increased pressure.

72. *Effect of heat.* It is found that if the temperature be increased, the elastic force of a quantity of air or gas which cannot change its volume is increased, but that if the air can expand, while its pressure remains the same, its volume will be increased.

To illustrate this, imagine an air-tight piston in a vertical cylinder containing air, and let it be in equilibrium, the weight of the piston being supported by the cushion of air beneath.

Raise the temperature of the air in the cylinder; the piston will then rise, or, if it be not allowed to rise, the force required to keep it down will increase with the increase of temperature.

73. *Thermometer.* As a general rule bodies expand under the action of heat, and contract under that of cold, and the only method of measuring temperatures is by observing the extent of the expansion or contraction of some known substance.

For all ordinary temperatures mercury is employed, but for very high temperatures a solid metal of some sort is the most useful, and for very low temperatures, at which mercury freezes, alcohol must be employed.

74. *The Mercurial Thermometer* is formed of a thin glass tube terminating in a bulb, and having its upper end hermetically sealed. The bulb contains mercury which also extends partly up the tube, and the space between the mercury and the top of the tube is a vacuum.

It must be observed that, as the glass expands with an increase of temperature, as well as the mercury, the apparent expansion is the difference between the actual expansion and the expansion of the glass.

In the Centigrade Thermometer the freezing point is marked 0⁰, and the boiling point 100⁰, the space between being divided into 100 equal parts, called degrees.

In Fahrenheit's Thermometer the freezing point is marked 32⁰, and the boiling point 212⁰; and in Reaumur's the freezing point is 0⁰, and the boiling point 80⁰.

75. *To compare the scales of these Thermometers.*

Let C, F and R be the numbers of degrees marking the same temperature on the respective thermometers; then, since the space between the boiling and freezing points must in each case be divided in the same proportion by the mark of any given temperature, we must have

$$C : F - 32 : R :: 100 : 180 : 80$$
$$:: 5 : 9 : 4,$$

or

$$\frac{C}{5} = \frac{F - 32}{9} = \frac{R}{4},$$

it being taken for granted that the temperatures indicated by the boiling point and the freezing point are the same in all.

The method of filling the thermometer, and the definitions of the freezing and boiling points, will be given at the end of the chapter.

76. *Pressure of the Atmosphere. Torricelli's Experiment.*

The action of the atmosphere was distinctly ascertained by the experiment of Torricelli. Taking a glass tube *AB*, 32 or more inches in length, open at the end *A* and closed at the end *B*, he filled it with mercury, and then, closing the end *A*, inverted the tube, immersed the end *A* in a cup of mercury, and then opened the end *A*. The mercury was observed to descend through a certain space, leaving a vacuum at the top of the tube, but resting with its surface at a height of about 29 or 30 inches above the surface of the mercury in the cup.

It thus appears that the atmospheric pressure, acting on the surface of the mercury in the cup, and transmitted, as we have shewn that such pressures must be transmitted, supports the column of mercury in the tube, and provides us with the means of directly measuring the amount of the atmospheric pressure.

In fact, the weight of the column of mercury in the tube above the surface in the cup, is exactly equivalent to the atmospheric pressure on an area equal to that of the section of the tube. This is about 15 lbs. weight on a square inch.

77. *Air has weight.* This may be directly proved by weighing a flask filled with air; and afterwards weighing it, when the air has been withdrawn by means of an air-pump. The difference of the weights is the weight of the air contained by the flask.

We are now in a position to account for the fact of atmospheric pressure. The earth is surrounded by a quantity of air, the height of which is limited, as may be proved by dynamical and other considerations; and if, above any horizontal area, we suppose a cylindrical column extending to the surface of the atmosphere, the weight of the column of air must be entirely supported by the horizontal area upon which it rests, and the pressure upon the area is therefore equal to the weight of the column of air.

According to this theory the pressure of the air must diminish as the height above the earth's surface increases, and, from experiments in balloons, and in mountain ascents, this is found to be the case. As before, taking Π for the pressure at any given place, and w as the intrinsic weight of the air, the pressure at the height z will be

$$\Pi - wz,$$

if we assume that the density, and therefore the intrinsic weight of the air, is sensibly the same through the height z.

It should be noticed that Π, measured in lbs. weight, is the pressure upon the unit of area at the place.

78. It has been mentioned that the pressure of a gas depends chiefly upon its volume and temperature, but it is implied in that statement that the gas is confined within a limited space, for without such a restriction the effect of its elasticity might be the unlimited expansion and ultimate dispersion of the gas.

The action of gravity is equivalent to the effect of a compression of the gas, and it is thus seen that the pressure of a gas is in fact due to its weight, as in the case of a liquid.

It may be shewn in the same manner as for air that any other gas has weight, and that the intrinsic weight is in general different for different gases.

Carbonic acid gas, for instance, is heavier than air, and this is illustrated by the fact that it can be poured, as if it were liquid, from one jar to another.

The Barometer.

79. This instrument, which is employed for measuring the pressure of the atmosphere, consists of a bent tube ABC, closed at A, and having the end C open.

The height of the portion AB is usually about 32 or 33 inches, and the portion BC is generally for convenience of much larger diameter than AB. The tube contains a quantity of mercury, and the portion AP above the mercury is a vacuum.

If the plane of the surface in BC intersect AB in Q, it is clear, since the pressure

at all points of a horizontal plane is the same, that the
pressure at Q is the same as the atmospheric pressure,
which is transmitted from the surface at C to Q, and
therefore the atmospheric pressure supports the column of
mercury PQ. Hence the height of this column is a measure
of the atmospheric pressure, and if w be the intrinsic weight
of mercury, and Π the atmospheric pressure,

$$\Pi = w \cdot PQ.$$

The density of mercury diminishes with an increase of
temperature, and it is an experimental result that, for an
increase of $1°$ centigrade, the expansion of mercury is
$\dfrac{1}{5550}$th, or $\cdot 00018018$ of its volume; and therefore if σ_t be
the density at a temperature t, and σ_0 at a temperature $0°$,

$$\sigma_0 = \sigma_t (1 + \cdot 00018018t),$$

or, if $\theta = \cdot 00018018$, $\sigma_0 = \sigma_t (1 + \theta t).$

Hence, since the intrinsic weights are proportional to the
densities,

$$w_0 = w_t (1 + \theta t),$$

and therefore, if Π is the pressure when the temperature is t
and when PQ is the height of the barometric column,

$$\Pi = w_t \cdot PQ = w_0 (1 - \theta t) PQ,$$

w_0 and w_t being the intrinsic weights of mercury at the
temperatures $0°$ and $t°$.

80. The *average height* of the barometric column at
the level of the sea is found to vary with the latitude, but
it is generally between $29\frac{1}{2}$ and 30 inches. This height is
however subject to continuous variations; during any one
day there is an oscillation in the column, and the mean
height for one day is itself subject to an annual oscillation,
independently of irregular and rapid oscillations due to high
winds and stormy weather. Usually the height of the
column is a maximum about 9 in the morning; it then
descends until 3 P.M., and again attains a maximum at 9 in
the evening.

81. *The Water Barometer.* Any kind of liquid will serve to measure the atmospheric pressure, but the great density of mercury renders it the most convenient for the purpose. If water were employed, it would be necessary to have a tube of great length; in fact, as the density of mercury is about 13·568 times that of water, the height of the column of water would be about $33\frac{1}{2}$ feet.

82. *Graduation of the Barometer.* Suppose the column of mercury to rise above P (fig. Art. 79); then it is clear that it descends below C in BC, and that the variation in the height of the column is the ·sum of these two changes.

Let k, K be the sectional areas of the tubes, and x the ascent above P, or the apparent variation; then the descent below C is $\dfrac{kx}{K}$, and the true variation is

$$x + \frac{kx}{K}, \text{ or } \left(1 + \frac{k}{K} \right) x.$$

Hence in graduation the distances actually measured from the zero point must be marked larger in the ratio of

$$1 + \frac{k}{K} : 1.$$

Again, since mercury expands rapidly with an increase of temperature, and since the scale on which the graduations are marked also expands, but not to the same extent, it is necessary to reduce the reading of the barometer to what it would be at some standard temperature.

This is usually taken to be the freezing point.

Let t be the temperature and h the observed height of the barometer.

Also take x to represent the fractional part of a volume of mercury which must be added to its volume for every degree of increase of temperature, and take y to represent the fractional part of its length by which the scale increases for each degree of temperature, the values of x and y being calculated for the particular thermometer in use.

Then the height which would have been observed had

the mercury been at the freezing point will be given by the formula,

for the Centigrade $\quad h - ht\,(x - y),$

and for Fahrenheit $\quad h - h\,(t - 32)\,(x - y).$

There is, further, a correction to be made for capillarity, the effect of which is to make the circle of contact of the surface of the mercury with the glass lower than it would be if the surface were flat, instead of being convex, as it really is.

83. *To find the atmosphere pressure on a square inch.*

This we can determine at once by observing that it is the weight of a cylindrical column of mercury of which the base is a square inch and the height equal to that of the barometric column.

The specific gravity of mercury is 13·568 times that of water; hence the atmospheric pressure on a square inch, taking 30 inches as the height of the barometer at the sea level,

$$= 30 \times 13\!\cdot\!568 \times 1000 \div 1728 \text{ oz. wt.}$$

$$= 14\!\cdot\!7 \text{ lbs. wt.}$$

This pressure varies from time to time, but is generally between $14\frac{1}{2}$ and 15 lbs. wt.

Taking the latter value, the pressure on a square foot would be equal to the weight of 19 cwt. 32 lbs.

84. *The height of the homogeneous atmosphere.*

If the density of the atmosphere were the same throughout the whole vertical column as it is at the sea level, its height would be less than 5 miles.

To prove this, let σ, ρ be the densities of mercury and of air; then, if h be the height of the barometer, the height of the atmospheric column would be $\dfrac{\sigma}{\rho}\,h$. Now, it has been found that at the level of the sea, the ratio $\sigma : \rho$ is about 10462 : 1, and if we take h to be 30 inches, we shall find that the height is a little less than 5 miles.

85. *The pressure of a given quantity of air, at a given temperature, varies inversely as the space it occupies.*

The experimental proof of this law, due to Boyle and Marriotte, is as follows.

A bent glass tube, the shorter branch of which can have its end closed, is fixed to a graduated stand. Both ends being open, a little mercury is poured in, which rests with its surfaces P, P in the same horizontal plane. The end A is now closed and more mercury is poured in at B; the effect is a compression of the air in AP, the mercury rising to a height Q, which is however below the surface R of the mercury in BP.

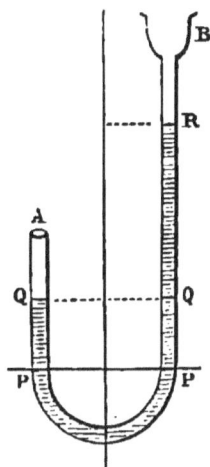

After closing the end A the pressure of the air is equal to the atmospheric pressure, and when more mercury has been poured in, the pressure of the air in AQ is equal to that of the mercury at Q, the same level in the longer branch. This latter pressure is due to atmospheric pressure on the surface R, and the weight of the column RQ.

If now the spaces AQ, AP be compared, which may be effected by comparing the weights of the mercury they would contain, and if the height h of the barometer be observed, it will be found that space AP : space AQ :: $h+QR:h$.

But, taking Π as the original pressure of the air in AP, and Π' as its pressure when compressed, and w as the intrinsic weight of mercury,

$$\Pi = wh, \text{ and } \Pi' = \Pi + wQR = w(h+QR);$$
$$\therefore \Pi' : \Pi :: \text{space } AP : \text{space } AQ,$$

and this proves the law for a compression of air.

For a dilatation, employ a bent glass tube of which both branches are long, and pour in mercury to a height P; then close the end A, and withdraw some of the mercury from the branch B.

If Q and R are the new surfaces it will be found that

$$\text{space } AP : \text{space } AQ :: h - QR : h;$$

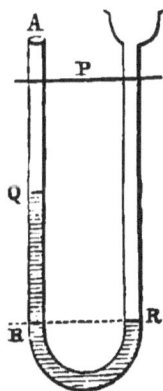

but, if Π'' is the pressure of the air when dilated,

$$\Pi'' = \text{Pressure at } R - wQR = w\,(h - QR),$$

and $\therefore \Pi'' : \Pi :: \text{space } AP : \text{space } AQ.$

In each case care must be taken to have the temperatures the same at the beginning and at the conclusion of the experiment.

Hence it follows that if p and p' are the pressures of a given mass of gas when its volumes are respectively V and V',

$$p : p' :: V' : V,$$

and therefore that, so long as the temperature is unchanged, pV is constant.

Also since ρV and $\rho'V'$ equally represent the mass of the gas, it follows that

$$p : p' :: \rho : \rho',$$

or that, for the same temperatures the pressure varies directly as the density, a law which may be otherwise expressed by means of the equation

$$p = k\rho.$$

86. Now the density of a given mass of gas at a given volume is a quantity which is independent of time and place, whereas p, representing the number of units of force exerted by the gas upon an unit of area, is a quantity, the numerical value of which depends upon locality.

It therefore follows that the value of k is dependent upon locality.

Taking a foot as the unit of length, and ρ pounds as the mass of a cubic foot of air close to the ground, the pressure on a square foot is equal to the weight, at the place, of $k\rho$ pounds, and therefore it follows that k is the height, in feet, at the place, of the homogeneous atmosphere.

87. *Effect of a change of temperature.*

If the pressure remain constant, an increase of temperature of $1°$ *centigrade, produces in a given mass of air an expansion* ·003665 *of what its volume would be at* $0°$ *centigrade under the given pressure*[*].

* This law was first published by Dalton in 1801, and by Gay-Lussac in 1802, independently of Dalton. It appears however that it had been obtained, some years before, by Charles.

This experimental law, combined with the preceding, enables us to express the relation between the pressure, density, and temperature of a given mass of air or gas.

Imagine a quantity of air confined in a cylinder by a piston to which a given force is applied, and let the temperature be 0° C. Raise the temperature to t^0: the piston will then be forced out until the original volume (V_0) is increased by ·003665 t . V_0 or $\alpha t V_0$, designating the decimal by α. If V be the new volume, we have

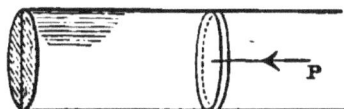

$$V = V_0 (1 + \alpha t),$$

and therefore, if ρ, ρ_0 be the densities at the temperatures t^0, 0^0, $\qquad \rho_0 = \rho (1 + \alpha t).$

Hence, $\qquad p = k\rho_0 = k\rho (1 + \alpha t).$

88. *Absolute Temperature.*

If we can imagine the temperature of a gas lowered until its pressure vanishes, without any change of volume, we arrive at what is called the absolute zero of temperature.

Assuming t_0 to be this temperature on the centigrade scale, we have $\qquad 1 + \alpha t_0 = 0,$

or $\qquad t_0 = -273^0.$

In Fahrenheit's scale this is -459^0.

Hence $\qquad p = k\rho (1 + \alpha t) = k\rho\alpha (t - t_0) = k\rho\alpha T,$

if T be the absolute temperature.

Taking V as the volume of the gas, ρV is constant, and therefore $\dfrac{pV}{T}$ is constant; *i.e. the product of the pressure and volume is proportional to the absolute temperature.*

The air Thermometer is a long straight tube of uniform bore closed at its lower end, open at the upper end, and containing air or some other gas, which is separated from the external air by a short column of liquid.

This thermometer is very sensitive, but it has the disadvantage that, as the atmospheric pressure is variable, no estimation can be

made of the temperature without at the same time taking account of
the height of the barometer*.

89.   *Illustration.*   The effect of heat in the expansion of
air may be illustrated by a simple experiment.

Take a glass tube, open at one end, and ending in a bulb
at the other; immerse the open end in
water, and then apply the heat of a
lamp to the bulb.   The air in the bulb
will expand, and will drive out a portion of the water in the tube, and may
drive out some air.

If the lamp be removed, the air
within will be cooled, and the water
will then rise in the tube to the same
level as before, or to a higher level.

90.   *Determination of heights by the barometer.*

It is found both from theory and from observation, that
the height of the barometric column depends on its altitude
above the sea level, and we are thus provided with a means
of directly inferring from observation the height of any given
station above the level of the sea.

For this purpose it is necessary to construct a formula
which shall connect the height of the barometer with the
height of its position above a given level, such as the sea
level.

A general formula would be somewhat complicated, and
difficult to obtain without the aid of the Integral Calculus,
since the atmospheric pressure depends on the temperature
and density of the air, which both vary with the height, and
also on the intensity of gravity, which diminishes with an
increase of height.

We shall however construct a formula on the supposition
that the temperature and the force of gravity remain constant: this will be practically useful for the determination of
comparatively small differences of altitude.

* See Chapter II. of Maxwell's Heat.

91. *If a series of heights be taken in arithmetic progression, the densities of the air decrease in geometric progression.*

Take a vertical column of the atmosphere of a given height z, and let it be divided into n horizontal layers of the same thickness, i.e. $\dfrac{z}{n}$, and suppose that ρ_1, ρ_2, ρ_3...ρ_n represent the densities of the successive layers, measuring upwards.

These layers may be supposed each of the same density throughout, and, if we take the temperature the same in all, the pressures on the upper sides of the layers will be $k\rho_1$, $k\rho_2$,...$k\rho_n$, k being the constant of variation, for the particular temperature, of the place.

The difference between any two consecutive pressures must be equal to the weight of the air between them, and if the horizontal section of the column is the unit of area these pressures are $k\rho_{r-1}$ and $k\rho_r$.

Hence, taking the unit of force to be the weight of a pound at the place, so that the numerical measures of the density and the intrinsic weight are the same, we have the equation

$$k\rho_{r-1} - k\rho_r = \rho_r \frac{z}{n}:$$

$$\therefore \frac{\rho_{r-1}}{\rho_r} = 1 + \frac{z}{kn},$$

that is, the densities diminish in geometric progression.

92. *To find an expression for the difference of the altitudes of two stations.*

If z be this difference, we have from the preceding article, putting γ for

$$1 + \frac{z}{kn},$$

and ρ_0 for the density immediately beneath the lowest layer of air,

$$\rho_{n-1} = \gamma \rho_n, \quad \rho_{n-2} = \gamma \rho_{n-1} \cdots \rho_1 = \gamma \rho_2, \quad \rho_0 = \gamma \rho_1,$$

and therefore, $\qquad\qquad \rho_0 = \gamma^n \rho_n.$

Hence, if p', p be the corresponding pressures
$$p = \gamma^n p'.$$
Let h', h be the observed altitudes of the barometer at the higher and lower stations respectively.

Then
$$\frac{h}{h'} = \frac{p}{p'} = \gamma^n = \left(1 + \frac{z}{kn}\right)^n.$$

And
$$\log_e \frac{h}{h'} = n \log \left(1 + \frac{z}{kn}\right)$$
$$= n \left(\frac{z}{kn} - \frac{1}{2}\frac{z^2}{k^2 n^2} + \dots\right)$$
$$= \left(\frac{z}{k} - \frac{1}{2}\frac{z^2}{k^2 n} + \dots\right).$$

Now the larger we make n, the more nearly our hypothetical case approaches to the continuous variation of the actual density of the air, and by taking n very large, we obtain the approximate expression,
$$z = k \log \frac{h}{h'},$$
observing that h' is less than h, that the temperature and the force of gravity are supposed constant throughout the height z, and that the numerical value assigned to k is its value at the place at which the observations are taken *.

The Siphon.

93.　The action of a siphon is an important practical illustration of atmospheric pressure.

---

* In the preceding article we have, for the sake of simplicity, taken the unit of force to be the weight of one pound at the place. If we had taken the unit of force to be the weight of a pound at some standard place, we should have had to introduce a symbol μ, to express the number of units of force in the weight of a pound at the place, and we should also have had to employ the value of k, say k', at the standard place. In that case we should have obtained
$$\gamma = 1 + \frac{\mu z}{k'n}, \text{ and } z = \frac{k'}{\mu} \log \frac{h}{h'}.$$
But, since kp and $k'p$ represent the same actual force in different units, it follows that
$$k' : k :: \mu : 1,$$
and therefore the final result is the same as that given above.

It is simply a bent tube ABC, which is open at both ends. When filled with water, the ends are closed and the siphon is then inverted, and one end C placed in water, the other end A being below the level of the surface of the water.

If the end C be opened, it is clear that the pressure at A is greater than the pressure at Q, which is equal to the pressure at P, and therefore to the atmospheric pressure.

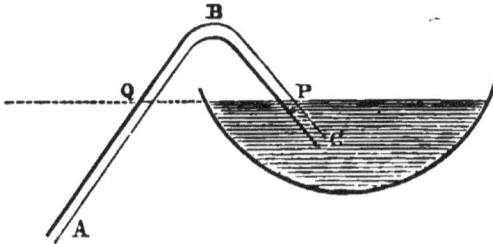

Hence, if the end A be unclosed, the water at A will begin to flow out, and by so doing diminish the pressure in the tube, and tend to form a vacuum in the upper portion of the tube. But if the height of B above the surface of the water be less than the height h of the water-barometer, the atmospheric pressure will force the water up the tube, and maintain a continuous flow through the end A, until either the surface has fallen below C, or, if the siphon be long enough, until it has descended so far that its depth below B is greater than h.

94. *Methods of filling and graduating a Thermometer.*

To fill the Thermometer with mercury a paper funnel is fastened to the open end, and mercury poured into it; the bulb is then heated over a spirit-lamp, a portion of the air in the tube is thereby expelled, and if the bulb be cooled the mercury descends in the tube. This process is repeated until the air is completely expelled, and when the tube is quite full and the mercury overflowing, the upper end is hermetically sealed by means of a blow-pipe; during the subsequent cooling the mercury contracts and descends, leaving a vacuum at the top of the tube *.

---

* This so-called vacuum is filled with the vapour of mercury.

The freezing and boiling points are now to be determined.

The freezing point is obtained by immersing the bulb and the lower portion of the tube in melting snow, and marking the tube outside at the end of the mercurial column.

The boiling point is obtained by immersing the bulb in the vapour of water boiling under a given atmospheric pressure, and marking the tube as before.

The temperature of steam depends on the atmospheric pressure, and it is therefore necessary to fix on some standard pressure, and to define the boiling point as the temperature of steam at that pressure. A barometric column of 30 inches at the level of the sea is the usual standard.

For the Centigrade Thermometer, the boiling point, 100°, is the temperature of steam when the height of the barometric column is 29·9218 inches (760 mm.), at the level of the sea in latitude 45°.

For some time after boiling the height of the mercury at the freezing temperature is gradually increased, and it has been found that it takes 4 or 5 years for the zero to attain its permanent position after boiling.

95. *Use of the Mercurial Thermometer limited.*

Mercury freezes at a temperature of $-40°$ C., and boils at a temperature of about 350° C.; it is therefore necessary for very high or very low temperatures to employ different substances.

For very low temperatures spirit of wine is used, and this liquid is generally employed in the construction of *minimum* Thermometers.

High temperatures are compared by observing the expansion of bars of metal or other solid substances, and various instruments, called *pyrometers*, have been constructed for this purpose.

96. *The differential Thermometer* is constructed in two different forms. In one form, of which the figure is a section, a horizontal tube branches upwards into two short vertical tubes ending in bulbs of equal size.

These bulbs contain air, and in the horizontal tube is a small portion of some coloured liquid, by which the air in one bulb is separated from the air in the other. The quantities of air are equal, so that when the bulbs have the same temperature the bubble of liquid rests at the middle of the tube: if however the temperatures be different, the liquid will rest in a position nearer to the bulb of lower temperature than to the other, since the air-pressure within it will be less than that in the other.

In the other form of the differential thermometer, the vertical portions, A, B, of the tube extend to a much greater height, and the liquid fills the whole of the horizontal portion

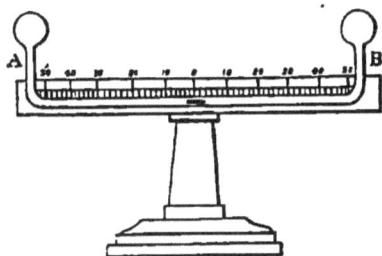

of the tube, and also partly fills the vertical portion of the tube.

The principle of the construction is the same, and the difference consists in the graduation of the vertical portions, instead of the horizontal portion of the tube.

On account of their great sensibility these thermometers are extremely useful in detecting small differences of temperature.

In graduating the second of these instruments, allowance must be made for the weight of the liquid, which is contained in the vertical tubes.

97. **Ex. 1.** *The same quantities of atmospheric air are contained in two hollow spheres; the internal radii being* r, r' *and the temperatures* t, t' *respectively, compare the whole pressures on the surfaces.*

Taking ρ, ρ' as the densities, we have, since the masses are equal, and the volumes in the ratio of $r^3 : r'^3$,

$$\rho r^3 = \rho' r'^3.$$

If p, p' be the corresponding pressures,

$$p = k\rho(1 + at), \quad p' = k\rho'(1 + at'),$$

and the pressures on the surfaces are

$$4\pi r^2 p, \text{ and } 4\pi r'^2 p',$$

which are in the ratio

$$r^2 \rho \,(1+at) : r'^2 \rho' \,(1+at'),$$

or $\qquad\qquad r' \,(1+at) : r \,(1+at').$

Ex. 2. *A hollow cylinder, open at the top, is inverted, and partly immersed in water; it is required to find the height of the surface of the water within the cylinder.*

Take b for the length of the cylinder, and a for the length not immersed.

Let x be the depth of the surface within below the surface without, Π, Π' the pressures of the
atmospheric air and of the compressed air in EC.
Then

$$\Pi' : \Pi :: b : a+x, \text{ Art. 83,}$$

and $\Pi' =$ pressure of the water at
the level $C = \Pi + wx$;

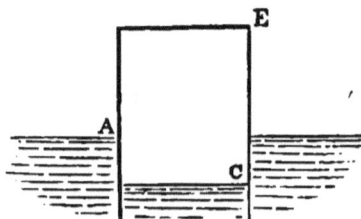

$$\therefore \frac{\Pi + wx}{\Pi} = \frac{b}{a+x}.$$

If h be the height of the water-barometer, $\Pi = wh$, and

$$\frac{h+x}{h} = \frac{b}{a+x},$$

or $\qquad\qquad x^2 + (a+h)\, x = (b-a)\, h.$

This equation gives two values for x, one positive and the other negative, the positive value being the one which belongs to the problem before us. The negative value is the result of another problem, the algebraical statement of which leads to the same quadratic equation.

Ex. 3. *A small quantity of air is left in the upper part of a barometer-tube; it is required to determine the effect on the height of the column.*

Let a be the length of the upper part of the tube which the air would occupy if its density were the same as that of the external air, and x the space it actually occupies, when the height of a true barometer is h.

If Π be the pressure of the external air, and Π' of the air in the space x,

$$\frac{\Pi'}{\Pi} = \frac{a}{x}.$$

Let h' be the height of the faulty barometer, then

$$\Pi = wh, \text{ and } \Pi' + wh' = \Pi \; ;$$

$$\therefore \frac{h-h'}{h} = \frac{a}{x} \text{ or } \frac{h'}{h} = 1 - \frac{a}{x} \dots (1).$$

The column is therefore depressed $\dfrac{ah}{x}$

or $\qquad\qquad \dfrac{ah'}{x-a}$ inches by (1).

Hence, if a be known, and h' and x be observed, the height of a true barometer can be inferred.

If a be unknown, it can be found from the equation (1) by taking simultaneous observations of h', x, and the height h of a true barometer.

EXAMINATION ON CHAPTER V.

1. If Fahrenheit's Thermometer mark 40°, what are the corresponding marks of Réaumur's and the Centigrade ?

2. When the mercurial barometer stands at 30 inches, what is the height of the barometer formed of a liquid of which the specific gravity is 5·6 ?

3. The air contained in a cubical vessel, the edge of which is one foot, is compressed into a cubical vessel of which the edge is one inch ; compare the pressures on a side of each vessel.

4. The air in a spherical globe, one foot in diameter, is compressed into another globe, 6 inches in diameter, and the temperature is raised by $t°$; compare the pressures of the air under the two conditions. Also compare the pressures on the surfaces of the globes.

5. What would be the effect of making a small aperture at the highest point of a siphon ?

6. If a barometer be held in a position not vertical, what would be the effect on the length of the column of mercury ?

7. If the sum of the readings on Fahrenheit's and the centigrade thermometer be zero for the same temperature, find the reading of each thermometer.

8. At the top of a mountain the barometer stands at 25 inches ; what would be the effect on the action of a siphon carried to the top ?

9. A siphon is filled with mercury, and held with its legs pointing downwards, and the ends closed ; what will be the effect of opening the ends, 1st, when they are, and 2ndly, when they are not, in the same horizontal plane ?

10. A cylindrical vessel contains water; how will a change in the height of the barometer affect the pressures on the base and curved surfaces of the cylinder, and to what extent?

11. A block of wood weighs, in air, exactly the same as a block of iron; which is really the heavier?

12. Examine the effects of making a small aperture, 1st, in the longer branch, 2ndly, in the shorter branch of the tube of a barometer?

13. Explain the uses, 1st, of the small hole which is made in the lid of a teapot, 2ndly, of a vent-peg.

14. Supposing the air half exhausted in a pair of Magdeburgh hemispheres, $1\frac{1}{2}$ ft. in diameter, find the force required to separate them, taking 15 lbs. weight as the atmospheric pressure on a square inch.

15. If a piece of glass float in the mercury within a barometer, will the mercury stand higher or lower in consequence?

16. Will any change in the action of a siphon be in any case coincident with a fall in the barometer?

17. A weight, suspended by a string from a fixed point, is partially immersed in water; will the tension of the string be increased or diminished as the barometer rises?

18. A mass of air at temperature 50^0 C. and pressure $33\frac{1}{4}$ inches of mercury, is compressed until its density is $\frac{4}{5}$ths of what it was before, its temperature at the same time falling to 16^0C.; find the new pressure.

19. If a given body lose in air, when h is the height of the barometer, the mth part of its weight, find what part of its weight it will lose when h' is the height of the barometer.

20. Find the greatest height over which a liquid of specific gravity $6\cdot784$ can be carried by means of a siphon when the barometer is at 30 inches.

21. A cannon is in the form of a cylinder 10 feet long. The powder takes up 4 inches of this, and the pressure where the powder is exploded is 10 tons on the square inch. Find, neglecting changes of temperature, what will be the pressure as the ball is leaving the muzzle.

22. A certain volume of gas at 0° C. is raised to 10° and its pressure is increased in the ratio $1\cdot03665$ to 1; in what ratio is this pressure further increased by an additional rise of 90° of temperature?

23. When the barometer column is 760 millimetres, a litre of dry air, at 0° C., contains $1\cdot293187$ grammes. Find the number of grammes in V litres of dry air at t° C., when h millimetres is the height of the barometer.

24. At a depth of 10 feet in a pond the volume of an air-bubble is ·0001 of a cubic inch; find approximately what it will be when it reaches the surface, if the height of the barometer is 30 inches, and the specific gravity of mercury 13·5.

25. If the specific gravity of air is ·0013, and that of mercury 13·568, and if the height of the barometer is 30 inches, prove that, a foot being the unit of length, the value of k is very nearly 26092.

NOTES ON CHAPTER V.

Thermometers were first constructed about the end of the sixteenth century, but the name of the inventor is not certainly known.

The various scales were formed in the early part of the 18th century; Fahrenheit's in 1714, at Dantzic ; Réaumur's in 1731 ; and the Centigrade by Celsius, a Swede, somewhat later.

The Aneroid Barometer. This instrument was invented by Vidi, and is exceedingly useful in mountain ascents on account of its small size and weight. Its construction depends on the varying effect of the atmospheric pressure on a thin metallic plate closing an exhausted chamber. A small metallic chamber, cylindrical in form, about an inch in height, and 2 or 3 inches in diameter, and closed by an elastic metal plate, is exhausted ; this is placed in a larger cylinder and the top of the elastic plate is connected by a system of levers with the hand of a graduated dial-face, so that any slight change of elevation or depression at the centre of the metallic plate is magnified and rendered visible by the motion of the hand.

Bourdon's Metallic Barometer, invented in 1850, is another instrument of a similar kind*.

It consists of an elastic flattened tube, ABC, of metal, exhausted of air, and bent very nearly into a circular form ; the middle part B is fixed and the rest of the tube is free. The section of the tube is like an ellipse, D, and it is found that if the atmospheric pressure increase, the tube becomes more curved, and the ends A, C approach each other ; and if it diminish, that the ends A, C separate. Hence if these ends be connected with the hand of a dial-face, the motion of the hand will mark the changes of atmospheric pressure.

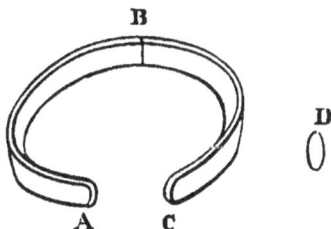

* The term Aneroid is sometimes applied to this instrument.

If the tube ABC, instead of being a vacuum, be connected by a pipe with the boiler of a steam-engine, or with any vessel containing air or gas, it becomes a very convenient *manometer* (see Art. 114), and is in fact sometimes used for this purpose on the engines of locomotives.

The Siphon. The general use of the siphon is to transfer liquids from one vessel to another without moving either vessel. It is useful in many other operations, such as draining a flooded field; and some time ago large siphons, 140 feet in length and $3\frac{1}{2}$ feet in diameter, were constructed for the purpose of draining the lands flooded by the inundation which occurred during the year 1862 on the eastern coast. These siphons were set working successfully. *The Times*, Oct. 1, 1862.

The Magdeburgh Hemispheres. A practical demonstration of the fact of atmospheric pressure was given by Otto von Guericke in 1654, who constructed this apparatus.

It consists of two hollow hemispheres of brass, fitting each other very accurately. A tube out of one of the hemispheres is screwed on the plate of an air-pump, and, when the two have been fitted together and the air exhausted, the stop-cock is turned, the apparatus removed from the air-pump, and a handle screwed on. Supposing the diameter of the hemispheres to be 3 or 4 inches, it will be found that a force of from 100 to 180 lbs. wt. will be necessary to separate them. The inventor employed hemispheres of nearly a foot in diameter, and shewed that a strain of more than 1500 lbs. wt. was required to force them asunder.

Taking the diameter as one foot, we can calculate the requisite force. The resultant pressure on one hemisphere is equal to the air-pressure on a circle one foot in diameter, that is, upon an area of 36π square inches. Making allowance for the fact that a perfect vacuum cannot be obtained, we may take 14 lbs. as approximately the pressure on a square inch, and the pressure is 504π or nearly 1583 lbs. wt.

Weight of the Air. Galileo measured the weight of the air by filling a globe with compressed air, and then weighing the globe. He employed a syringe to force the air into the globe; and, in order to find the quantity of air, he placed the globe in an inverted glass receiver filled with water, then opened it, and observed the amount of water displaced.

Torricelli and Pascal. The experiment of Torricelli, described in Art. (73), was made in the year 1643, one year after the death of Galileo, who had remarked the fact that a pump would not raise water to a greater height than 32 or 33 feet, but was unable to account for it. It was reserved for his pupil and successor, Torricelli, to explain the real cause of the phenomenon, and his experiment was repeated and its consequences were extended by Pascal a few years later.

Torricelli shewed that the pressure of the air supports the column of mercury in a barometric tube; Pascal demonstrated that the weight of the air is the cause of the pressure. Amongst various experiments, Pascal had a water-barometer constructed, but his most valuable idea was a suggestion that the heights of a barometer, at the foot and at the top of a mountain, should be compared. This was effected by his friend Perier in 1648, who ascended the Puy de Dome in Auvergne, and ascertained the fact of a fall of nearly 4 inches in the barometer at the top of the mountain. The observations were repeated in various ways, on the roofs of houses, and in cellars, and it was thus rendered clear that the weight of the air is the immediate cause of the existence of the barometric column.

The two treatises of Pascal, *De l'équilibre des liqueurs, et de la pesanteur de la masse de l'air*, contain the theory of the pressure of fluids, and give complete explanations of the actions of siphons and pumps, and of many common phenomena ; the main object however of these treatises is to demonstrate the unphilosophical character of the old explanation that the abhorrence of nature to a vacuum accounted for the rise of water in a pump, and that this abhorrence did not exist beyond a rise of 32 feet.

It appears that Descartes was acquainted with the fact that air has weight, and indeed he made a suggestion that the reason why water will not rise beyond a certain height is *the weight of the water which counterbalances that of the air*.

Balloon Ascents. The fall of the barometer in balloon ascents is a means of determining the altitude attained.

In a balloon ascent by De Luc, the barometer at the greatest height stood at 12 inches ; but in a later balloon ascent by Mr Glaisher, the column was seen to descend to less than 10 inches, implying a height of nearly six miles ; and it is probable, as the observations were interrupted by the severity of the cold, and the rarity of the air, that an altitude of more than six miles was attained. *The Times*, Sept. 9, 1862.

EXAMPLES.

1. THE temperature of the air in an extensible spherical envelope is gradually raised $t°$, and the envelope is allowed to expand till its radius is n times its original length ; compare the pressure of the air in the two cases.

2. A volume of air of any magnitude, free from the action of force, and of variable temperature, is at rest : if the temperature at a series of points within it be in arithmetical progression, prove that the densities at these points are in harmonical progression.

3. A given mass of elastic fluid of uniform temperature is confined in a smooth vertical cylinder by a piston of given weight ; neglecting the weight of the fluid, shew how to find its volume.

4. A piston moves freely in a closed air-tight cylinder, the axis of which is vertical. When the piston is in the middle of the cylinder, the air above and the air below are of the same density. Find the position of equilibrium of the piston.

5. A vertical closed cylinder is half filled with water, the other half being occupied by air of a given density and temperature; if the temperature be raised $t°$, find the increase of the whole pressure on the base, and on the curved surface of the cylinder.

6. If h, h' be the heights of the surface of the mercury in the tube of a barometer above the surface of mercury in the cistern at two different times, compare the densities of the air at those times, the temperature being supposed unaltered.

7. A vertical cylinder, containing air, is closed by a piston, which is tied by an elastic string fastened to its central point, and also to the base of the cylinder. If when the piston is in equilibrium the string have its natural length, determine the effect on the length of the string of increasing the temperature of the air in the cylinder by a given number of degrees.

8. If under an exhausted receiver a cylinder sinks to a depth equal to three-fourths of its axis; find the alteration in the depth of immersion when the air (specific gravity $= ·0013$) is admitted.

9. A body is floating in a fluid; a hollow vessel is inverted over it and depressed: what effect will be produced in the position of the body, (1) with reference to the surface of the fluid within the vessel, (2) with reference to the surface of the fluid outside?

10. A pipe 15 feet long, closed at the upper extremity, is placed vertically in a tank of the same height; the tank is then filled with water; shew that, if the height of the water-barometer be 33 feet 9 inches, the water will rise 3 feet 9 inches in the pipe.

11. A vessel, in the form of a prism, whose base is a regular hexagon, is filled with air; prove that, if every rectangular face of the prism be capable of turning freely about its edges, and the prism be then compressed so that its base becomes an equilateral triangle, the pressure of the air within it will be increased in the ratio of 3 to 2.

12. A conical wine-glass is immersed, mouth downwards, in water; how far must it be depressed in order that the water within the glass may rise half way up it?

13. A jar contains water in which a hollow rigid envelope open at the bottom and partially filled with air just floats; the top of the jar is closed by an elastic membrane, and a small space between it and the water is filled with air; on pressing the membrane inwards the envelope sinks; explain this.

14. A barometer is held suspended in a vessel of water by a string attached to its upper end, so that a portion of the string is immersed; find the height of the mercury and the tension of the string.

15. A piston, the weight of which is equal to the atmospheric pressure on one of its ends, is placed in the middle of a hollow cylinder which it exactly fits, so as to leave a length a at each end filled with atmospheric air. The ends of the cylinder are then closed, and the cylinder is placed with its axis inclined at an angle a to the vertical; shew that the piston will rest at a distance $a\{(1+\sec^2 a)^{\frac{1}{2}} - \sec a\}$ from its former position.

16. A cylinder, open at both ends, is half immersed in water, its axis being vertical; the upper end is then closed, and the cylinder is raised until its lower end is very near the surface of the water outside; find the height to which the water rises inside.

17. Two barometers of the same length and transverse section each contain a small quantity of air; their readings at one time are h, k, and at another time h', k'; compare the quantities of air in them.

18. A glass cylinder, 1 square inch in section, is filled with water and inverted over a trough, so that its closed top stands 5 feet above the level of the trough. 20 cubic inches of gas at 87° C. are allowed to bubble into the cylinder, and to cool to 15° C. the temperature of the room. What volume does the gas now occupy, the height of the water-barometer being 32 feet? (Coefficient of dilatation of gas $= \frac{1}{273}$ per degree centigrade.)

19. A quantity of gas contained in a sphere is compressed into the cube which can be inscribed in the sphere; compare the whole pressure on the surface of the cube and the sphere.

If gas in a cubical vessel is compressed into the sphere which can be inscribed in the cube, the whole pressures on the two surfaces are equal.

20. A right cylindrical vessel on a plane base contains a quantity of gas, which is confined within it by a disc exactly similar and parallel to the base; prove that the pressure on the curved surface of the cylinder is independent of the position of the disc.

21. Assuming that a change from 30 inches to 27 inches in the height of the barometer corresponds to an altitude of 2700 feet, find the altitude corresponding to the height 21·87 inches of the barometer.

22. Having given that the specific gravity of air and mercury are respectively ·00129 and 13·596, and the height of the barometer 75·9 centimetres, prove that if the unit of force be taken to be the weight of 800 kilogrammes, the numerical value of the pressure of the air will be almost exactly equal to that of its density, it being assumed that the mass of a cubic centimetre of water is one gramme.

23. Having given the equation $pr = RT$, calculate in Fahrenheit's scale of temperature the numerical value of R for a pound of air, supposing that a cubic foot of air is ·0763 of a pound at a temperature of 62° F. when the barometer is 30 inches high; taking an inch as the unit of length, the weight of a pound as the unit of force, the density of mercury relative to water as 13·596, and −273° C. as the absolute zero of temperature.

24. The readings of a faulty barometer containing some air are 29·4 and 29·9 inches, the corresponding readings of a correct instrument being 29·8 and 30·4 inches respectively; prove that the length of the tube occupied by the air is 2·9 inches when the reading of the barometer is 29 inches, and find the corresponding correct reading.

25. If the height of the barometer vary from one end of a lake to the other, shew that there will be a heaping up of the water on one side. Find what will be the greatest rise above the mean level produced in a circular lake of 100 miles diameter by a variation in the height of the mercury of ·001 inch per mile.

26. A barometer tube consists of three parts whose sections starting from the lowest are A, B, C. The column consists partly of mercury and partly of glycerine, so that for a certain atmospheric pressure the glycerine just fills that part of the tube whose section is B. Shew that if $A : B :: B : C :: 1 : \lambda$, and if μ is the ratio of the density of glycerine to that of mercury, the sensitiveness of this barometer is greater than that of a mercury barometer in the ratio $1 : \lambda + \mu - \lambda\mu$, the alteration of level in the cistern being neglected.

CHAPTER VI.

THE DIVING-BELL, COMMON PUMP, LIFTING PUMP, FORCING
PUMP, FIRE-ENGINE, BRAMAH'S PRESS, AIR-PUMPS, BARO-
METER GAUGE, SIPHON GAUGE, CONDENSER, MANOMETERS,
BARKER'S MILL, PIEZOMETER, HYDRAULIC RAM, AND
STEAM-ENGINE.

The Diving-Bell.

98. THIS is a large bell-shaped vessel made of iron, open
at the bottom, and containing seats for several persons. Its
weight is greater than that of the water it would contain,
and, when lowered by a chain into the water, the air within
it is compressed, but will prevent the water from rising high
in the bell, and the persons seated within are thus enabled
to descend in safety to considerable depths.

When the surface of the water within the bell is at a
depth of 33 feet below the outer surface the bell will be
half filled with water, and the compression of the air would
of course increase with the depth, but the difficulty arising
from this compression is overcome by forcing fresh air from
above through a flexible tube opening under the mouth of
the bell. There are also contrivances for the expulsion of
the air when rendered impure.

Tension of the Chain. This is equal to the weight of
the bell diminished by the weight of water displaced by
the bell and the air within. It is therefore evident that
unless fresh air is forced in from above the tension of the
chain will increase as the bell descends.

99. *Supposing the bell cylindrical, and that no air is supplied from above, it is required to find the height to which the water rises in the bell.*

If the bell be partially immersed, we fall upon a case already considered, Ex. 2. Ch. v.

If the bell be wholly immersed, let b represent the length of the cylinder, a the depth of its top, and x the length occupied by air.

The pressure of the air within $= \Pi \dfrac{b}{x}$

$$= \Pi + w(a+x);$$

and $\therefore$ if $\quad \Pi = wh,$

$$hb = (h+a)x + x^2,$$

and as before the positive value of x is the one required.

If A be the area of the top of the bell, and if we neglect its thickness, the volume of water displaced is Ax, and the tension of the chain

$$= \text{weight of bell} - wAx.$$

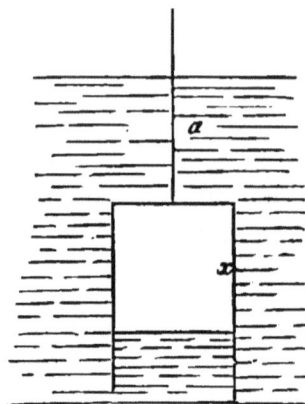

The Common Pump.

100. The Pump most commonly in use is a Suction-pump, of which the figure is a vertical section.

AB, BC are two cylinders having a common axis, M is a piston moveable over the space AB by means of a vertical rod, connected with a handle, D is a spout a little above A, and C the surface of the water in which the lower part of the pump is immersed: also in the piston, and at B, are valves opening upwards.

Action of the Pump. Suppose the piston at B and the pump filled with ordinary atmospheric air; raising the piston, the air in BC will open the valve B, and then, expanding as the piston rises, its pressure will be less than that of the atmosphere at C outside the pump; hence the atmospheric

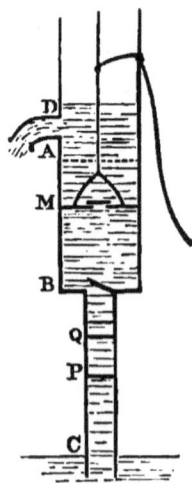

pressure on the surface of the water outside will force water up the tube BC, until the pressure at C is equal to the atmospheric pressure.

As the piston rises the water will rise in BC, the pressure of the air above M keeping the valve M closed. When the piston descends, the valve B closes, and the air in MB becoming compressed will open the valve M, and escape through it.

This process being repeated, the water will at length ascend through the valve B, and at the next descent of the piston will be forced through the valve M and be then lifted to the spout D, through which it will flow.

The height BC must be less than the height (h) of the water-barometer, or else the water will never rise to the valve B.

It is not essential to the construction that there should be two cylinders; a single cylinder, with a valve somewhere below the lowest point of the piston-range will be sufficient, provided the lowest point of the range be less than 33 feet above the surface in the reservoir.

In each case the height above the water in the reservoir of the piston-range should be considerably less than 33 feet; otherwise the quantity of water lifted by the piston at each stroke will be small.

In the figure the tubes are represented as straight tubes; this is not necessary to the working of a pump, and the tube below the piston-range may be of any shape, and may enter the reservoir at any horizontal distance from the upper portion of the pump.

101. *Tension of the Piston-rod.* If the water in BC has risen to P when the piston is at M, the pressure Π' of the air in MP = pressure of water at P = pressure at $C - w.PC$

$$= \Pi - w.PC.$$

But if A be the area of the piston, the tension of the rod is the difference between the atmospheric pressure above and the pressure $\Pi'A$ below, i. e. ($\Pi - \Pi'$) A, or $wPC.A$.

If one inch be taken as the unit of length, and if h be the height in inches of the water barometer, wh is approximately equal to the weight of 15 lbs., and therefore the tension is approximately equal to the weight of

$$15 \frac{PC}{h} A \text{ lbs.}$$

102. *To find the height through which the water rises during one stroke of the piston.*

Let P and Q be the surfaces of the water at the beginning and end of an upward stroke of the piston, that is, while the piston is raised from B to A.

The air which at the beginning of the stroke occupied the space BP occupies at the end of it the space AQ; but the pressures are respectively, if $\Pi = wh$,

$$w(h - PC), \; w(h - QC).$$

Hence　　　　$h - PC : h - QC :: \text{vol. } AQ : \text{vol. } BP.$

If r, R be the radii of the cylinders (Fig. Art. 100),

$$\text{vol. } AQ = \pi R^2 . AB + \pi r^2 . BQ = \pi R^2 AB + \pi r^2 (BC - QC),$$
$$\text{vol. } BP = \pi r^2 . BP = \pi r^2 (BC - PC),$$
$$\frac{h - PC}{h - QC} = \frac{R^2 . AB + r^2 (BC - QC)}{r^2 (BC - PC)},$$

and for any given value of PC this equation determines QC.

103. If the range of the piston be less than AB, as for instance AE, then EC must be less than h. Moreover, a limitation exists with regard to the position of E.

For, if P be the surface of the water when the piston M is at A, then as the piston descends, the valve B will close, but the valve M will not be opened until the pressure of the air in MB is greater than the atmospheric pressure.

When M is at A the pressure of the air $= w(h - PC)$, and, unless the valve is opened before M arrives at E, the pressure of the air in

$$EB = w(h - PC) \frac{AB}{EB},$$

which must be greater than wh, and therefore $h . AE$ must be greater than $AB . PC$. Hence, to ensure the opening of the valve while the surface is below B, we must have

$$h . AE > AB . BC;$$

i.e. AE must be at least the same fraction of AB that BC is of h.

This condition, although in all cases necessary, may not be sufficient.

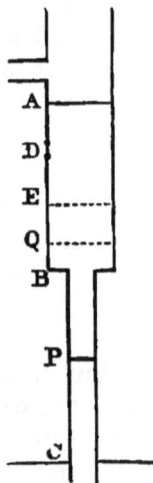

For, suppose that when M is at A, the surface of the water is at Q, in which case the pressure of the air in $AQ = w\,(h - QC)$.

When the piston descends to E, the pressure in EQ

$$= w\,(h - QC)\frac{AQ}{EQ},$$

which must be greater than wh,

and $\therefore h\,.\,AE > AQ\,.\,QC.$

The greatest value of $AQ\,.\,QC$ is $\frac{1}{4}AC^2$, and $\therefore$ we must have

$$h\,.\,AE > \frac{1}{4}AC^2.$$

Since $\frac{1}{4}AC^2 > AB\,.\,BC$, unless B is the middle point of AC,

it follows that this latter condition includes the preceding, which is therefore in general insufficient.

These conditions must be also satisfied in the case of the pump with a single cylinder.

104. *Tension of the rod when the pump is in full action.*

In the figure of the previous Article, let $CD = h$; then it will be seen that, at each stroke, the volume DE of water is lifted, and therefore the tension of the rod when the piston is ascending will be $wA\,(h + ED)$ until the water begins to flow through the spout.

If A be on a level with the spout, all the water lifted will be discharged, and, as the piston descends, the tension of the rod will be wAh.

The Lifting Pump.

105. By means of this instrument, water can be *lifted* to any height. It consists of two cylinders, in the upper of which a piston M is moveable; the piston-rod works through an air-tight collar, and a valve opens outwards at D leading into a vertical tube. When the piston ascends, lifting water, the valve D opens and water ascends in the tube; when the piston descends the valve D closes, and every successive stroke increases the quantity of water in the tube. The only limitation to the height to which water can be lifted is that which depends on the strength of the instrument, and the power by which the piston is raised.

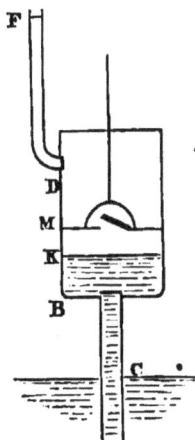

Tension of the rod. If $CK = h$ the piston lifts the volume BK at each stroke, and, as the air is expelled before the machine is in full action, the tension $= wA \cdot KB$, until the water is lifted to the valve D. The power applied to the piston-rod must be then increased until the pressure of the water opens the valve D, that is, until the pressure $= w(h + FD)$, F being the surface of the water in the tube. The water will then be forced up the tube, the tension of the rod increasing as the surface F ascends.

The Forcing Pump.

106. In this pump the piston M is solid, and ranges over the space AE. At B and D are valves opening upwards, DF being a tube leading out of AB.

When this pump is first set in action, it works as a common pump, the air at each descent of the piston being driven through D, and the water rising in BC. When however the water has risen through B, the piston, descending, forces it through D, and when the piston ascends, the valve D closes and more water rises through B. The next descent forces more water through D, and it is obvious that water can be thus forced upwards to any height consistent with the strength of the instrument.

The stream which flows from the top of the tube will be intermittent, but a continuous stream can be obtained by employing a strong air-vessel DL, out of which the vertical tube passes upwards. The air in the upper part of the vessel is condensed, and exerts a varying, but continuous pressure on the surface of the water within the vessel, and if the size of the vessel be suitable to that of the pump, and to the rate of working it, the air pressure will not have lost its force before a new compression is applied to it, and thus a continuous, although varying, flow will be maintained.

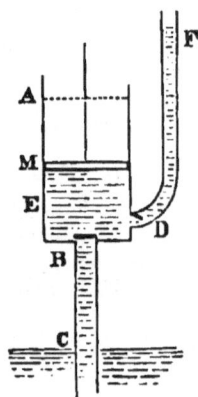

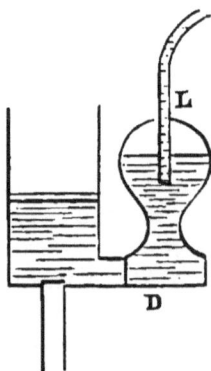

The Fire Engine.

107. The Fire Engine is only a modification of the Forcing-pump with an air-vessel, as just described.

Two cylinders are connected with the air-vessel, and the pistons are worked by means of a lever GEG', so that while one ascends the other descends. The vertical tube out of the air-vessel has a flexible tube of leather attached to it, by means of which the stream can be thrown in any direction.

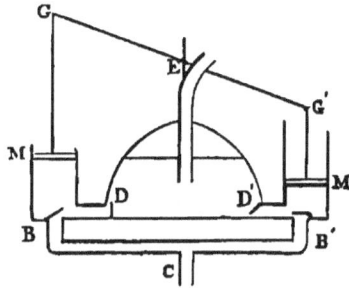

Bramah's Press.

108. This instrument is a practical application of the principle of the transmission of fluid pressures.

In the figure, which represents a vertical section of the instrument, A and C are two solid cylinders working in air-tight collars; EB and FD are strong hollow cylinders connected by a pipe BD; at B is a valve opening inwards, and at D a valve opening upwards, a pipe from D communi-

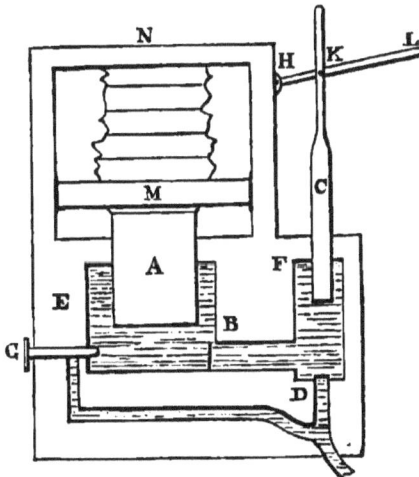

cating with a reservoir of water. *M* is a moveable platform, on which the substance to be pressed is placed, and *N* is the top of a strong frame; *HKL* is the lever working the cylinder *C*, *H* being the fulcrum, and *L* the handle.

Action of the Press. Suppose the spaces *EB*, *FD* filled with water, and *C* in its lowest position; on raising *C*, the atmospheric pressure forces water from the reservoir into *FD*, and when *C* is afterwards forced down, the valve *D* closes, the valve *B* is opened, a portion of the water in *FD* is driven into *EB*, and the cylinder *A* is then made to ascend. A continued repetition of this process will produce any required compression of the substance between *M* and *N*.

At *G* there is a plug which can be unscrewed when the compression is completed.

The Force produced. If *P* be the power applied at the handle *L*, the force on *C* downwards is $P\dfrac{HL}{HK}$. Let *r*, *R* be the radii of the cylinders *C* and *A*, and *p* the pressure of the water,

then
$$\pi r^2 p = P\frac{HL}{HK},$$

and the pressure on $A = \pi R^2 p = P\dfrac{HL}{HK}\cdot\dfrac{R^2}{r^2}.$

It is obvious that by increasing the ratio of *R* to *r*, any amount of pressure may be produced.

We have taken for granted in describing the action of the press that the cylinders at first were full of water. If this is not the case the water will be pumped up from the reservoir by the action of the cylinder *C*, and whatever air there may be within will be compressed until its pressure is the same as that of the water.

Presses of this kind were employed in lifting into its place the Britannia Bridge over the Menai Straits.

109. The portion *C* of the instrument is sometimes called a *Plunger Pole Pump*, and an important part of the machine is the construction of the water-tight collars at *E* and *F*, as without these water under great pressure would

force its way between the pole and the hollow cylinder in which it works.

A circular aperture DE is made in the side of the cylinder, and a piece of leather is doubled over a metal ring within it. The figure is a vertical section of the cylinder and collar, and it will be seen that the water pressing on the under side of the leather keeps it in close contact with the side of the cylinder, and the greater the pressure the closer the contact, so that no escape of water can possibly take place, · unless the leather be torn.

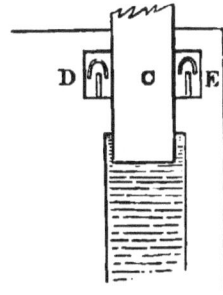

•

Hawksbee's Air-Pump.

110. Two cylinders, AB, $A'B'$, are connected by pipes leading from B and B' through C with a receiver. Pistons MM' are worked in the cylinders by means of a toothed wheel, and at B, B' and in the pistons are valves opening upwards.

Suppose M at its highest and M' at its lowest position, and turn the wheel so that M descends and M' ascends; the valve B closes and the air in MB being compressed flows through the valve M, while the valve M' closes, and air from the receiver flows through B' into $M'B'$.

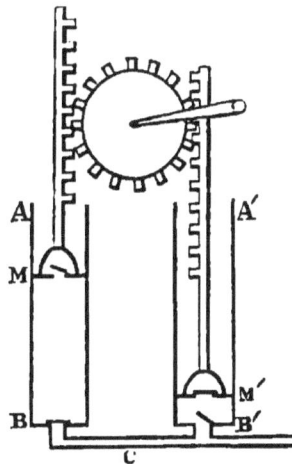

When the wheel is turned and M' descends, the valve B' closes and the air in $M'B'$ flows through M', while the valve M closes and air from the receiver flows through B. At every stroke of the piston a portion of the air in the receiver is withdrawn, and it is evident that a degree of exhaustion may be thus obtained, limited only by the weight of the valves which must be lifted by the pressure of the air beneath.

Let A be the volume of the receiver, and B of either

cylinder; ρ the density of atmospheric air and $\rho_1, \rho_2, \ldots \rho_n$ the densities in the receiver after $1, 2, \ldots n$ descents of the pistons.

After the first stroke the air which occupied the space A will occupy the space $A + B$, and therefore

$$\rho_1 (A + B) = \rho A ;$$

similarly $\rho_2 (A + B) = \rho_1 A$;

$$\therefore \rho_2 (A + B)^2 = \rho A^2,$$

and after n strokes

$$\rho_n (A + B)^n = \rho A^n.$$

Hence if Π_n be the pressure of the air in the receiver after n strokes, and Π of the atmospheric air,

$$\frac{\Pi_n}{\Pi} = \frac{\rho_n}{\rho} = \left(\frac{A}{A + B} \right)^n.$$

In working the instrument, the force required is that which will overcome the friction, together with the difference of the pressures on the under surfaces of the pistons, the pressures on their upper surfaces being the same.

It will be seen that a perfect vacuum cannot be obtained by this instrument, but, since the density decreases in geometric progression as the number of strokes increases, a very large proportion of the air can be withdrawn if the instrument be constructed with sufficient care.

Smeaton's Air-Pump.

111. This instrument consists of a cylinder AB in which a piston is worked by a rod passing through an air-tight collar at the top; a tube from B leads into a glass receiver C, and at A and B, and in the piston there are valves opening upwards.

Supposing the receiver and cylinder to be filled with atmospheric air, and the piston at B; raising the piston, the air in AM is compressed, opens the valve A, and flows out through it, while at the same time a portion of

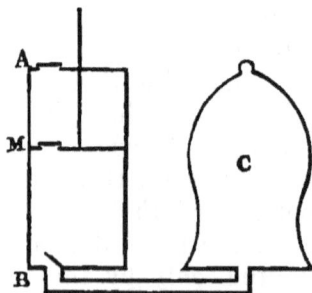

the air in C flows through the valve B, so that when the piston arrives at A, the air which at first occupied C now fills both the receiver and the cylinder. When the piston descends, the valves B and A close and the valve M opens; the air in AB passes above the piston, and as the piston rises is forced through A, which is opened as soon as the pressure in M becomes greater than the atmospheric pressure. Thus at every stroke a portion of the air in the receiver is forced out through A.

If ρ be the density of atmospheric air, ρ_n the density in the receiver after n strokes of the piston, and A, B the volumes of the receiver and cylinder respectively, then, as in the previous article,

$$\rho_n (A + B)^n = \rho A^n,$$

observing that the volume of the connecting tube is neglected.

An advantage of this instrument is that, the upper end of the cylinder being closed, when the piston descends, the valve A is closed by the external pressure, and the valve M is then opened easily by the air beneath. Moreover, the labour of working is diminished by the removal, during the greater part of the stroke, of the atmospheric pressure on M, which is only exerted while the valve A is open during the latter part of the ascent of the piston.

A greater degree of exhaustion may be obtained by making the B aperture in the side of the cylinder without a valve, and working the piston, a solid one with or without a valve, below the aperture B. The limitation arising from the weight of the valve at B is thus removed, and the only limitations left are those which arise from the weight of the valve at A, and the exact fitting of the piston and receiver.

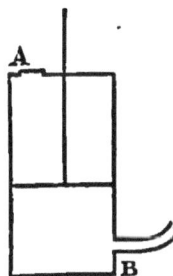

The Barometer Gauge.

112. The density of the air in the receiver of an air-pump at any moment is shewn by this instrument.

It is simply a barometric tube, the upper end of which communicates with the receiver, while the lower end is immersed in a cup of mercury, so that, as the pressure in the receiver diminishes, the mercury will rise in the tube.

If x be the altitude, PQ, of the mercury in the gauge, and h the height of the barometer, the pressure of the air in the receiver $= wh - wx$, if w be the intrinsic weight of mercury.

Hence the density in the receiver is to the density of atmospheric air $:: h - x : h$.

It is important to use this gauge for experiments requiring strict accuracy, but for less important experiments a siphon gauge may be used.

The Siphon Gauge.

113. This is a glass tube $ABCD$, the end D of which can be screwed on a pipe communicating with the receiver.

The end A is closed and the portion AB completely filled with mercury, which also fills a small part BP of BC.

If AP be not more than 28 inches in length, the tube AB will at first remain completely filled, but as the exhaustion proceeds, the mercury will sink in AB and rise in BC, and if at any time x be the difference of the heights in AB and BC, wx will be the pressure in the receiver, and the density will therefore be $\rho \dfrac{x}{h}$.

The Condenser.

114. This instrument is employed in the compression of air.

A hollow cylinder AB has one end screwed into the neck of a strong receiver C; at B is a valve opening inwards, and a piston M also has a valve opening inwards.

Suppose the cylinder and receiver filled with atmospheric air and the piston to be at A; forcing the piston down, the air in MB is compressed, and, opening the valve B, is forced into the receiver. When the piston is drawn back, the valve B is closed by the air in the receiver, and the valve M is opened by the outer air which flows in and fills the cylinder: this air is forced into the receiver at the next stroke, and at every succeeding stroke the same quantity of air is added to the receiver.

After n strokes, the volume of air of density ρ, forced into the receiver, is $A + nB$, A being the volume of the receiver and B of the cylinder; hence, if ρ_n be its density,

$$\rho_n A = \rho (A + nB), \text{ or } \frac{\rho_n}{\rho} = 1 + n\frac{B}{A}.$$

Gauge of a Condenser. A glass tube AB, closed at the end B, and connected with the condenser at the end A contains atmospheric air in the portion BC, which is separated from the air in the condenser by a drop of mer-

cury which rests at C before the compression commences. As the condensation proceeds, the drop of mercury is forced towards B, until the density in BC is the same as the density in the condenser. Thus when the mercury is at D the density $= \rho \dfrac{BC}{BD}$.

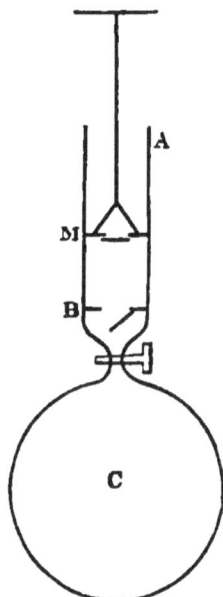

Sprengel's Air-Pump.

115. A glass tube BD, which is longer than a barometer tube, and is open at both ends, is fixed in a vertical position. A funnel is fitted closely to its upper end and the lower end dips into a glass vessel into which it is fixed by means of a cork. This vessel has a spout a little higher than the lower end of the tube. From the upper part of the tube a lateral tube C branches off and communicates with the receiver. Mercury being poured into the funnel, it runs down and closes the lower end of the tube so that no air can enter from below.

More mercury being poured into the funnel, the process of exhaustion begins, and the tube BD is seen to be filled with falling columns of mercury separated by columns of air. Air and mercury escape through the spout, and the mercury is collected in the basin E. This mercury can be poured back into the funnel, until the exhaustion is completed, and then the receiver may be closed.

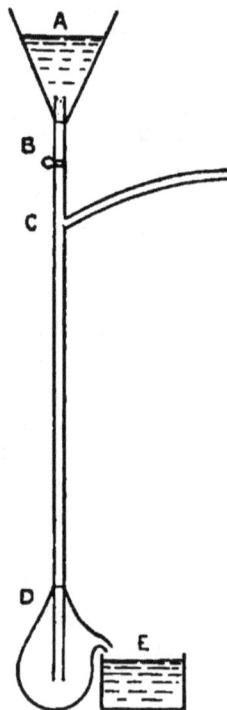

Manometer.

116. The term manometer is applied to any instrument for measuring the pressure of condensed air or gas of any kind, when its elastic force is greater than that of the atmosphere. The gauge of a condenser, for instance, is a manometer. The term however is sometimes applied to any instrument, such as the barometer-gauge, for measuring the elastic force of air or gas under any circumstances.

The annexed figure represents a manometer, the principle of which is nearly the same as that of the gauge of a condenser.

AB is a vertical glass tube, closed at the end A and containing dry air in the part AP; the tube ends in a strong bulb B containing mercury, and from this bulb a tube BC proceeds, leading to the vessel which contains the condensed air or gas. When the air in the tube C is ordinary atmospheric air at a given pressure, the mercury stands at the same level CC' in both tubes, but when the tube BC is connected with air or gas at a higher pressure the mercury rises in $C'A$, compressing the air above it, until the pressure in PA is equal to the pressure in EC diminished by the pressure due to the column PE' of mercury.

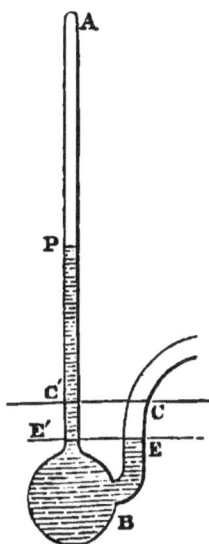

To find the relation between the pressure to be measured and the height of the mercury.

Let Π' represent the pressure in EC, and Π'' the pressure in PA ;

$$\text{then } \Pi''=\Pi \cdot \frac{AC'}{AP}, \text{ and } \Pi''+wPE'=\Pi' ;$$

$$\therefore \ \Pi'=w \cdot PE'+\Pi \frac{AC'}{AP}.$$

Let k, K be the sectional areas of the tubes AC', CE ;

$$\therefore \text{ if } PC'=x, \ CE=\frac{xk}{K}, \text{ and } PE'=x+\frac{xk}{K},$$

$$\therefore \Pi'=wx\left(1+\frac{k}{K}\right)+\Pi\frac{a}{a-x}, \text{ where } a=AC',$$

or, if $\Pi=wh$, $\Pi'=wh'$,

$$\frac{h'}{h}=\frac{x}{h}\left(1+\frac{k}{K}\right)+\frac{a}{a-x}.$$

This equation gives the ratio of the pressure required to the atmospheric pressure.

The graduation of the instrument depends on the solution of the equation ; thus, making $h'=2h$, $3h$, &c., the successive proper values of x mark the altitudes for pressures of 2, 3,...atmospheres.

117. The *Siphon Manometer* is a long glass tube ABC, open at the end A, and communicating at the end C with the gas or vapour, the pressure of which is to be measured.

The tube contains mercury, and the height of the mercury in AB above its equilibrium level measures the excess of the pressure in the part BC of the tube above the atmospheric pressure.

Then if the mercury ascend to P in AB, and descend to E in CB, CC' being the original level, $CE = C'P$, and therefore, if $C'P = x$, and $\Pi' =$ pressure in CB,

$$\Pi' = \Pi + w \cdot 2x,$$

or $\Pi' - \Pi \propto x.$

A graduated scale is attached to the tube AB, and, from the equation above, it is seen that the length of $C'P$ corresponding to a pressure of n atmospheres is $\dfrac{n-1}{2} h$, if h be the height of the barometer. Hence by giving successive integral or fractional values to n, the graduation of the scale can be effected.

The manometers we have now described are constructed on purely hydrostatic principles, but there are others, depending on different mechanical principles, and a very useful one, from its portability, is *Bourdon's Metallic Manometer*, which has the additional advantage of not being fragile. The construction of this instrument is briefly explained in the notes appended to Chapter v.

Barker's Mill.

118. ACB is a tube, capable of revolving about its axis which is vertical, and having two or more horizontal tubes BE, BD connected with it. C is a cup through which water can be poured down the tube, and at D and E, in the sides of BD and BE, orifices are made which open in opposite directions. Suppose a stream of water to flow into C and through the tubes; as the water flows through BD the

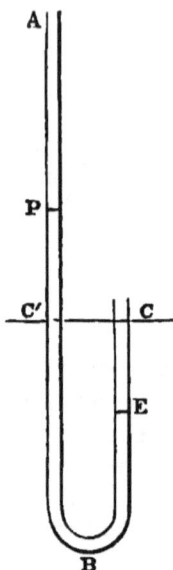

pressures on the sides balance each other except at D, at which part of the tube there is an uncompensated pressure on the side opposite the orifice, the effect of which is to turn the tube CD round. The same effect is produced by the

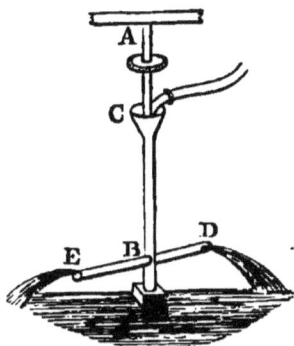

water issuing at E, and a continued rotation of the instrument is thus produced. By means of a toothed wheel at A the instrument may be employed in communicating motion to other machines, and in maintaining such motion.

The Piezometer.

119. This is an instrument for measuring the compressibility of liquids.

A thermometer tube CD, open at the end C, is enclosed in a strong glass vessel, which also contains a condenser-gauge EF. (See Art. 114.)

The liquid to be examined is poured into CD, and a drop of mercury is then introduced into CD so as to isolate the liquid, and the vessel is filled with water and closed by a piston. This piston A is moveable in the neck of the vessel, and, by means of a screw B, any required pressure can be produced. The gauge EF measures the pressures, and the compression of the liquid is obtained by observing the space through which the drop of mercury P is forced.

The area of a section of CD and the volume of the bulb are found by weighing the

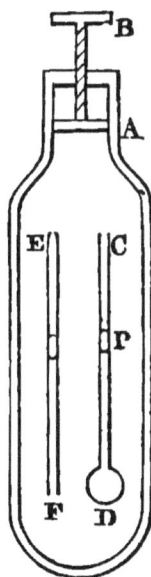

quantities of mercury contained by the bulb and a portion
of the tube.

The Hydraulic Ram.

120.   The fall of water from a small height produces
a momentum which by means of the Hydraulic Ram * is
utilized and made to produce the ascent of a column of
water to a much greater height.

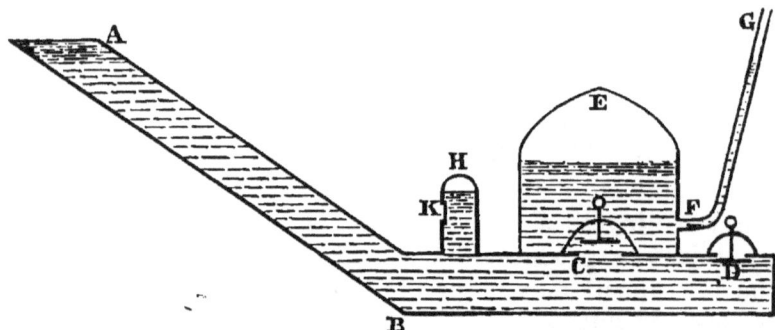

The figure is a vertical section of the machine, AB being
the descending and FG the ascending column of water,
which is supplied from a reservoir at A. E is an air-vessel
with a valve at C, opening upwards; at D is a valve opening
downwards, and H is a small auxiliary air-vessel with a valve
K opening inwards.

The action of the Machine.   The valve D will at first be
open in its lowest position, and if water descend from A, a
portion will flow through D, but the action on the valve will
soon close it, and the sudden check thus produced increases
the pressure; the valve C is lifted and water flows into the
vessel E, and condenses the air within; the reaction of the
air thus condensed forces water up the tube FG.

During this process the pressure of the water in the
large tube diminishes, and the valves C and D both fall;
the fall of the latter produces a rush of water through the
opening D, followed by an increased flow down AB, the
result of which is again the closing of D, and a repetition

---

* Invented by Montgolfier.

of the process just described, the water ascending higher in *FG*, and finally flowing through *G*.

The action of the machine is assisted by the air-vessel *H* in two ways, first, by the reaction of the air in *H* which is compressed by the descending water, and secondly by the valve *K* which affords supplies of fresh air. When the water rises through *C*, the air in *H* suddenly expands, and its pressure becoming less than that of the outer air, the valve *K* opens, and a supply flows in, which compensates for the loss of the air absorbed by the water and taken up the column *FG*, or wasted through *D*. About a third of the water employed is wasted, but the machine once set in motion will continue in action for a long time provided the supply in the reservoir be maintained.

The Atmospheric Steam-Engine.

121. This instrument, constructed by Newcomen soon after the year 1700, was the first in which the oscillation of a beam was maintained by the elastic force of steam.

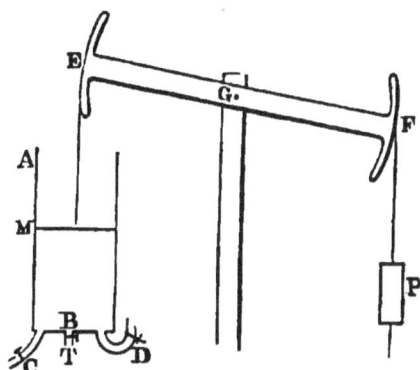

A solid beam *EGF*, which is moveable about *G*, has its ends arched; to these ends chains are attached which are connected with the rod of a piston in a cylinder *AB*, and with a rod supporting a weight *P*, this weight being less than the atmospheric pressure on the piston. *C* is a pipe connected with a boiler, *B* a pipe opening by a stop-cock, and *D* is a pipe connected with a cistern of cold water.

This engine was first used for working the pumps of mines, and a rod Q attached to P is connected with the piston-rod of a pump.

The stop-cocks at C and D are connected with the beam, so that when M is at A, C is closed, and D opens, and when M is at B, C opens and D is closed. The stop-cock at B is made to open when M descends to B, and to close immediately after.

Action of the Engine. The pressure of the steam in the boiler is a little greater than that of the atmosphere, and when M is at B, C is open, and steam rushes into MB; hence the weight P will cause the piston to ascend. When M reaches A, C is closed, D is opened, and a jet of cold water is thrown in, condensing the steam, and thereby producing very nearly a vacuum below M. The pressure of the air on the piston being greater than the weight P forces the piston down, and when it has descended, C again opens, and an oscillation of the piston is thus maintained.

As B opens when M descends to the lowest point of its range the water flows out before the ascent.

In the actual engine constructed by Newcomen the stop-cocks were turned by hand, but an attendant, left to work them, invented the machinery by which the engine became self-acting.

The Single-acting Steam-Engine.

122. In the atmospheric engine, the cooling of the cylinder at each stroke of the piston causes a great loss of power, for the steam on first entering the cylinder is partially condensed, and its elastic force is therefore diminished. One of Watt's first improvements was to produce the condensation in a separate vessel. The tube D was made to communicate with a vessel containing cold water, the space above the water being a vacuum. This vacuum could be produced by filling the vessel with steam and then condensing it by cooling the vessel. When the piston is at A, the stop-cock opens and the steam rushes into the vacuum, and is therefore condensed by the cold water. A pump from the condensing vessel was connected with the beam, so that

the overplus of water arising from the condensed steam would be drawn off as soon as formed. These two changes in the atmospheric engine constitute the single-acting engine, but the additional change of making the steam drive the piston downwards as well as upwards, leads to the double-acting engine, the type of most of the steam-engines now in actual use.

Watt's Double-acting Steam-Engine.

123. The cylinder AB, in which the piston works, is closed at both ends, the piston ranging from a to b. The end of the piston-rod is connected by means of a jointed parallelogram with the end E of the beam EGF, and the

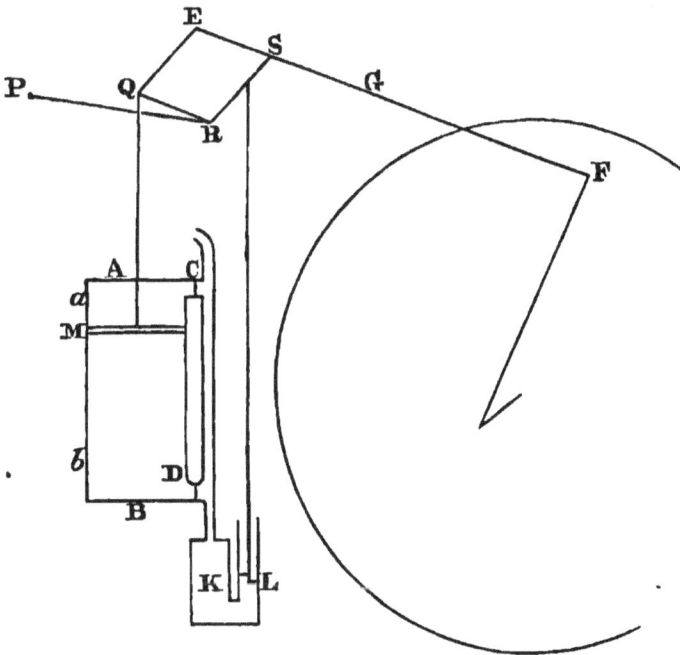

end F of the beam is attached to the crank of the fly-wheel. At C and D there are stop-cocks which are connected with the fly-wheel, so that when M arrives at a, the steam flows from the boiler through C into AM, and when M arrives at *b*, the steam flows through *D* into *BM*. In each case the steam is shut off when *M* has passed over about one-third of its range.

K, the condenser, is surrounded with cold water, and *L* is a pump connected with it; a tube from *K*, not drawn in the figure, is connected with *C* and *D* so that when steam from the boiler flows into *AM*, the steam from *MB* flows into *K*, and when steam from the boiler flows into *MB*, the steam from *AM* escapes into *K*.

Supposing *M* to be at *a*, steam enters *AM* from the boiler and forces the piston down, its expansive force being sufficient to complete the piston-range after it is cut off; on arriving at *b*, the steam in *AM* escapes into *K* and is condensed, and fresh steam from the boiler enters *MB*, drives the piston upwards, and then escapes into *K* and is condensed. The continued accumulation of water in *K* is prevented by the pump *L*, by which it is drawn off at every stroke.

The use of the fly-wheel is to maintain a continuous motion, and prevent the irregularity which would arise from the intermittent action of the piston.

Parallel motion. The parallelogram *EQRS* represents a system of jointed rods, invented by Watt for the purpose of making the end *Q* of the piston-rod move very nearly in a vertical line. The point *R* is connected with a fixed centre at *P*, and, by a proper adjustment of the lengths of the rods, it is found that the point *Q* deviates very slightly from the vertical during its motion.

A full account of the various contrivances for parallel motion will be found in Professor Willis's *Mechanism.*.

The High-Pressure Engine.

124. In the double-acting engine the pressure of the steam need not be greater than the atmospheric pressure. In the high-pressure engine it is many times greater, and the steam instead of being condensed is let off into the open air at each stroke. The condenser and air-pump are thus rendered unnecessary, and the engine simplified. The engines of locomotives on railways are high-pressure engines.

These descriptions give the main principles on which the construction of steam-engines depend, but for the various forms in which these principles are developed, and the innumerable details of the mechanism connected with them, the reader must consult special treatises on the subject, such as Dr Lardner's in Weale's series, Bourne's works on the Steam-engine, or the excellent article in the *Encyclopædia Britannica*.

EXAMINATION UPON CHAPTER VI.

1. A diving-bell is lowered until the surface of the water within is 66 feet below the outer surface; state approximately how much the air is compressed.

2. If a small hole be made in the top of a diving-bell, will the water flow in, or the air flow out?

3. To what height could mercury be raised by a pump?

4. In a Bramah's Press, HK is 1 inch, HL is 4 inches, the diameter of A is 4 inches, and that of C is half an inch; find the force on A produced by a force of 2 lbs. applied at L.

5. If the receiver be 4 times as large as the barrel of an air-pump, find after how many strokes the density of the air is diminished one half.

6. State any limitations which exist to the degree of exhaustion producible by an air-pump.

7. What must be the height of a Siphon Manometer that it may mark a pressure of 60 lbs. on a square inch?

8. The diameter of the piston of a Lifting pump is 1 foot, the piston-range is $2\frac{1}{2}$ feet, and it makes 8 strokes per minute; find the weight of water discharged per minute, supposing that the highest level of the piston-range is less than 33 feet above the surface in the reservoir, and that 33 feet is the height of the water-barometer.

9. If, in working the same pump, the lower level of the piston-range be $31\frac{1}{2}$ feet above the surface in the reservoir, find the weight discharged per minute.

10. In exhausting a receiver by an air-pump a cloud is sometimes seen in the receiver; explain the cause of this.

11. If the receiver and the barrel of an air-pump are in the proportion of 4 to 1, find how much has been pumped out at the end of the fifth stroke.

12. How would the tension of the rope of a diving-bell be affected by opening a bottle of soda-water in the bell?

13. If a cylindrical diving-bell, height 5 feet, be let down till the depth of its top is 19 feet, find the space occupied by the air, the water-barometer standing at 33 feet.

14. A person seated in a diving-bell, which descends slowly, observes an incompressible float on the level of the fluid within the bell. What change in the plane of floatation of the float takes place in the descent?

NOTES ON CHAPTER VI.

Archimedes' Screw.

This instrument, one of the earliest hydraulic machines on record, is employed for raising water, and depends for its action only on the weight and mobility of the particles of water.

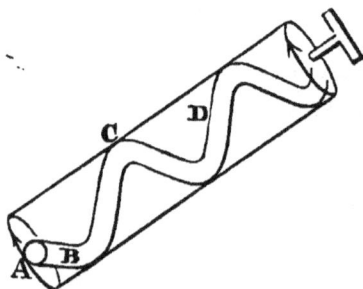

Let $ABCD$ be a metal tube, bent into the form of a corkscrew, and then held so that its axis is inclined to the vertical, and let it be moveable about its axis. The axis is to be inclined so much to the vertical, that a stone, inserted at A, will fall to B, and after oscillating rest at B. In the figure the tube is drawn as if wound round a cylinder moveable about its axis.

If we turn the cylinder in direction of the arrows, B will ascend, and the portions of the tube from B to C will successively take the same positions as B relative to the axis of the cylinder; as they do so, the stone at B will fall into those positions, and thus be gradually passed along the tube. Instead of the stone, suppose water poured in at A; the turning of the instrument will gradually raise the water until it flows out at the upper end. If the end A be immersed in water, a continued stream will ascend and flow out above.

Tradition assigns to Archimedes the credit of the invention of this instrument, and it is certain that its use dates at least as far back as the time of Archimedes. It was employed in Egypt in draining the land after an inundation of the Nile.

The point B at which the stone will rest is not underneath the cylinder but on one side, the ascending side, and between the middle and the under part of the surface of the cylinder : this can be seen experimentally.

Speaking strictly, the point B lies between the lowest generating line of the cylinder, and the generating line which lies halfway between the highest and lowest generating lines.

The machine will not act unless the inclination of the axis of the cylinder to the vertical be greater than the pitch of the screw, i. e. the inclination of the thread of the screw to a circular section of the cylinder. If these inclinations be equal, the point B is on the side of the cylinder, on the middle generating line, and the descending tangent BT is directed downwards at all other points. To make this clear, take a cylinder, of which BF is a diameter; let the dotted line represent a portion of the thread of a screw, BT being the tangent at B, and turn the cylinder round BF, which is supposed to be horizontal, until BT is horizontal : the inclination of the axis to the vertical is then equal to the pitch of the screw.

Turn the cylinder further, and if the screw mark the direction of a tube, it is an Archimedes' screw, in a position to work freely in raising water.

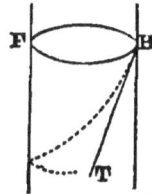

The Piezometer.

In the *Annales de Chimie et de Physique,* Vol. xxxi., 1851, a full account is given, by M. Grassi, of experiments with this instrument on the compressibility of water and some other liquids, and also on the compressibility of glass : these experiments were a continuation of M. Regnault's on the compressibility of water and mercury.

The apparatus employed by M. Grassi is identical in principle with the piezometer of the text, but differs in details. In one particular point the difference is of practical importance ; instead of producing pressure by a screw, the pressure on the surface of the water is produced by means of condensed air. The advantages gained are that the pressure can be measured with greater precision, and that it can be adjusted more easily, and changed more gradually.

The following are Grassi's final conclusions with regard to water :

(1) The compressibility of distilled water, deprived of air, varies with the temperature, and diminishes as the temperature increases.

(2) For distilled water, the compression due to one atmosphere is the same whatever be the pressure, provided the temperature remain constant.

EXAMPLES.

1.   If P be the weight of a diving-bell, P' of a mass of water the bulk of which is equal to that of the material of the bell, and W of a mass of water the bulk of which is equal to that of the interior of the bell, prove that, supposing the bell to be too light to sink without force, it will be in a position of unstable equilibrium, if pushed down until the pressure of the enclosed air is to that of the atmosphere as W to $P - P'$.

2.   After a very great number of strokes of the piston of an air-pump the mercury stands at 30 inches in the barometer-gauge, the capacity of the barrel being one-third that of the receiver, prove that after 3 strokes the height of the mercury is very nearly 17·34 inches.

3.   A fine tube of glass, closed at the upper end, is inverted, and its open end is immersed in a cup of mercury, within the receiver of a condenser; the length of the tube is 15 inches, and it is observed that after 3 descents of the piston the mercury has risen 5 inches; the height of the barometer being 30 inches, find how far it will have risen after four descents.

4.   A diving-bell is immersed in water so that its top is at a depth a below the surface, the height of the air in the bell being then x, and the height of the water-barometer h. If now a bucket of water of weight W be drawn up into the bell, shew that the tension of the chain is increased by $\dfrac{Wx}{h+a+2x}$.

5.   A diving-bell is suspended at a fixed depth; a man who has been seated in the bell suddenly falls into the water and floats. Determine the effects on (1) the tension of the chain, (2) the level of the water in the bell, (3) the amount of water in the bell.

6.   If a cylindrical diving-bell, whose capacity is V cubic feet, be sunk to such a depth that the water stands at $\dfrac{1}{m}$ th of its height, and be then lowered at the uniform rate of n feet per second, prove that the number of cubic feet of air at the atmospheric pressure which must be pumped in per second in order that the water may always remain at the same height will be $\left(1 - \dfrac{1}{m}\right) \dfrac{n}{h} V$, where h is the height of the water-barometer in feet.

7.   The length of the lower pipe of a common pump above the surface of the water is 10 feet, and the area of the upper pipe is 4 times that of the lower: taking 33 feet as the height of the water-barometer, prove that if at the end of the first stroke the water just rise into the upper pipe, the length of the stroke must be very nearly 3 feet 7 inches.

8. A cylindrical diving-bell, of height a, is furnished with a barometer and lowered into a fluid : the heights of the mercury in the barometer before and after immersion being h and h' respectively, shew that the depth of the bottom of the bell below the surface of the fluid is equal to $\left(\dfrac{\sigma}{\rho}+\dfrac{a}{h'}\right)(h'-h)$, where σ is the specific gravity of mercury, and ρ that of the fluid.

9. A bent tube, the arms of which are vertical, and which is open at one end and closed at the other, is partially filled with mercury, the density of the air between the mercury and the closed end of the tube being initially equal to that of the external air. If this tube be placed within the receiver of an air-pump, investigate a formula for determining the difference of heights of the mercury, in the two arms of the tube, after n strokes of the piston.

10. The valve in the piston of an air-pump being of given size and weight, find at what point of the nth descent the valve will be raised.

11. If h be the range of the piston of an air-pump, a its distance from the top of the barrel in its highest position, β its distance from the bottom in its lowest position, and ρ the density of the atmosphere ; prove that the limiting density of the air in the receiver will be

$$\frac{a\beta}{(h+a)(h+\beta)}\,\rho.$$

12. In the $\overline{n+1}|^{th}$ ascent of the piston of a Smeaton's air-pump, find the position of the piston when the highest valve (whose weight may be neglected) begins to open ; and shew that then the tension of the piston-rod : the pressure of the atmosphere on the piston

$$:: 1-\left(\frac{A}{A+B}\right)^{n} : 1-\left(\frac{A}{A+B}\right)^{n}\frac{B}{A+B}.$$

13. A cylindrical diving-bell of internal volume v, is filled with air at atmospheric pressure Π and absolute temperature T, and is lowered to a certain depth below the surface of water. Shew that if a small rise (x) in the temperature and increase (y) in the atmospheric pressure now take place, the apparent weight of the bell will be unaltered provided $\dfrac{x}{T}=\dfrac{yv'}{\Pi v}$, v' being the volume of the air in the bell.

14. Air is uniformly forced into a condenser, the condenser contains a gauge consisting of a drop of mercury C in a fine horizontal glass tube : if A be the position of the mercury when the air is uncompressed, B the end of the tube, prove that the ratio $AC : CB$ increases uniformly.

15. A condenser and a Smeaton's air-pump have equal barrels and the same receiver, the volume of either barrel being one-twentieth of

that of the receiver; shew that if the condenser be worked for 20 strokes, and then the pump for 14, the density of the air in the receiver will be approximately unaltered.

16. If a condenser be fitted with a gauge formed by a tube containing air which is separated from the air in the receiver by a drop of mercury, the distances the drop of mercury has moved from its initial position after 1, 2, 3...strokes are in harmonical progression.

If the piston do not reach to the bottom of the cylinder in which it works, shew that after n strokes the pressure in the receiver is

$$\Pi \frac{v+x}{x} - \left(\frac{V}{V+x}\right)^n \Pi \frac{v}{x},$$

where Π is the pressure of the atmosphere and v, $v+x$ are the least and greatest volumes between the piston and bottom of the cylinder and V is the volume of the receiver.

CHAPTER VII.

METHOD OF DETERMINING SPECIFIC GRAVITIES. SPECIFIC GRAVITIES OF AIR AND WATER, THE HYDROSTATIC BALANCE, THE COMMON HYDROMETER, SIKES'S, NICHOLSON'S, AND HARE'S HYDROMETERS, THE STEREOMETER.

To compare the specific gravities of air and water.

125.   TAKE a large flask, which can be completely closed by a stop-cock, and exhaust it by means of an air-pump.

Weigh the flask, and then permit the air to enter, and weigh the flask again.   Finally find the weight of the flask when filled with water.

Let w be the weight of the exhausted flask, w', w'' its weights when filled with air and water;

∴ $w' - w =$ weight of the air contained by the flask,

and $w'' - w =$ ....water.........................

Hence $w' - w$ and $w'' - w$ being the weights of equal volumes of air and water,

specific gravity of water : that of air :: $w'' - w : w' - w$.

In the same manner the specific gravity of any gas can be compared with that of water.

The specific gravity of water at $20\cdot5°$ is about 768 times that of air at $0°$ under the pressure of $29\cdot9$ inches of mercury at $0°$.

To compare the specific gravities of two fluids by weighing the same volume of each.

Let w be the weight of a flask, w' its weight when filled with one fluid (A), and w'' its weight when filled with the other fluid (B).

Then

$w' - w$ = weight of the fluid A contained in the flask,

$w'' - w$ = ...B...........................;

∴ specific gravity of A : that of B :: $w' - w$: $w'' - w$.

If the flask be not exhausted when its weight is determined, then, for strict accuracy, w must be diminished by the weight of the air which the flask contains.

126. *To find the specific gravity of a solid broken into small fragments.*

Put the broken pieces in a flask, fill the flask with water and let its weight be then w''; let w be the weight of the flask when filled with water, and w' the weight of the solid in air.

Then

$w'' - w$ = weight of solid pieces – weight of the water they displace ;

　　　 = w' – weight of water displaced ;

therefore

$w' + w - w''$ = weight of water displaced,

$$\text{and } \frac{\text{specific gravity of solid}}{\text{that of water}} = \frac{w'}{w' + w - w''}.$$

If we take account of the air displaced by the solid, its real weight is greater than w' by the weight of air displaced. This weight must therefore be added to w'.

The Hydrostatic Balance.

127. The hydrostatic balance is an ordinary balance, having one of the scale-pans smaller than the other, and at a less distance from the beam, so that weights immersed in water may be suspended from it.

The following cases are examples of its use.

(1) *To compare the specific gravities of a solid and a liquid.*

Let w be the weight of the solid in air.

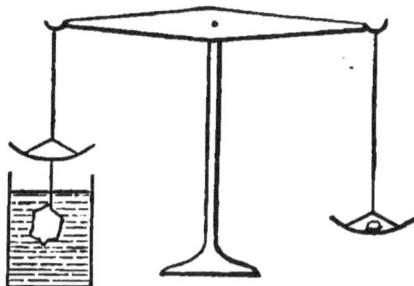

Place the liquid in a vessel, as in the figure, and suspend the solid from the scale-pan.

Let w' be the weight of the solid in the liquid,

$\therefore w - w'$ is the weight lost by the solid, and is therefore the weight of the liquid displaced by the solid, Art. (39);

and w, $w - w'$ are the weights of equal volumes of the solid and liquid.

Hence,

specific gravity of solid : that of liquid :: $w : w - w'$.

If we take account of the air displaced by the solid, we must add to w the weight of the air it displaces, since its true weight is diminished by exactly this weight of air.

This remark applies also to the next two articles.

128. We have tacitly supposed the solid to be specifically heavier than the liquid. If it be lighter it must be attached to a heavy body of sufficient size and weight to make the two together sink in the liquid.

Let $w =$ the weight of the solid in air,

$x =$ the weight in air of the heavy body attached to it,

$x' =$ the weight in the liquid of the heavy body,

$w' =$ the weight in the liquid of the two together.

$w + x - w' =$ the weight of liquid displaced by the two together, since it is the weight lost.

$x - x' =$ weight of liquid displaced by the heavy body.

Hence

$w + x' - w' =$ weight of liquid displaced by the solid,

and therefore $\dfrac{\text{specific gravity of solid}}{\text{specific gravity of liquid}} = \dfrac{w}{w + x' - w'}$.

129. (2) *To compare the specific gravities of two liquids.*

Take a solid which is specifically heavier than either liquid, and let w be its weight in air.

Let $w' =$ weight of solid in one liquid (A),

and $w'' = $ .... the other liquid (B);

$\therefore w - w' =$ weight of liquid A displaced by the solid,

$w - w'' =$ .... B .... ;

$\therefore$ specific gravity of A : that of B :: $w - w'$: $w - w''$.

The Common Hydrometer.

130. The common hydrometer consists of a straight stem ending in two hollow spheres B and C.

This hydrometer is usually made of glass, and the sphere C is loaded so that the instrument will float with the stem vertical.

When the hydrometer is immersed and allowed to float in a liquid, it displaces its own weight of the liquid, and by observing the positions of equilibrium in two liquids, the volumes displaced are inferred, and the specific gravities of the liquids can be compared.

Let κ be the area of a section of the stem, and v the volume of the instrument.

Suppose that when floating in a liquid (A) the level D of the stem is in the surface, and that in liquid (B) the level E is in the surface.

Then, since the specific gravities are in the ratio of the intrinsic weights, it follows that if s and s' are the specific gravities of A and B respectively,

$$s(v - \kappa . AD) = s'(v - \kappa . AE);$$

$$\therefore \frac{s}{s'} = \frac{v - \kappa . AE}{v - \kappa . AD}.$$

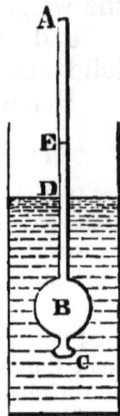

Sikes's Hydrometer.

131. This instrument differs from the common hydrometer in the shape of the stem, which is a flat bar and very thin, so that it is exceedingly sensitive. It is generally constructed of brass, and is accompanied by a series of small weights F, which can be slipped over the stem above C so as to rest on C.

The use of the weights is to compensate for the great sensitiveness of the instrument, which would without the weights render it applicable only to liquids of very nearly the same density.

Suppose the instrument floating in a liquid (A), with the level D of the stem in the surface, and that w' is the weight on C. In a liquid (B) let E be in the surface, and w'' the weight at C.

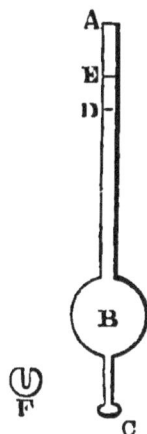

Let w be the weight of the instrument, v its volume, κ the section of the stem, v', v'' the volumes of w', w'', and s', s'' the specific gravities of the liquids.

Then $w + w' =$ weight of fluid A displaced,

$$v + v' - \kappa \cdot AD = \text{volume of } A \text{ displaced};$$

$$\therefore w + w' = s'\,(v + v' - \kappa \cdot AD) \times (62\cdot5) \text{ lbs. weight.}$$

Similarly

$$w + w'' = s''\,(v + v'' - \kappa \cdot AE) \times (62\cdot5) \text{ lbs. weight,}$$

and therefore $\dfrac{s'}{s''} = \dfrac{w + w'}{w + w''} \cdot \dfrac{v + v'' - \kappa \cdot AE}{v + v' - \kappa \cdot AD}$.

If the liquid (B) be the standard liquid, $s'' = 1$, and s', the specific gravity of (A) is at once determined.

Nicholson's Hydrometer.

132. The two hydrometers just described are used for comparing the specific gravities of fluids; Nicholson's hydrometer can be also employed in comparing the specific gravities of a solid and a fluid.

It consists of a hollow vessel B, generally of brass, supporting a cup A by a very thin stem, which is often
a steel wire, and having attached to it a heavy cup
C: on the stem connecting A and B a well-defined
mark D is made.

We proceed to explain the use of the instrument in the two cases.

(1) *To compare the specific gravities of two
liquids.*

If w be the weight of the hydrometer, w' the
weight which must be placed in A in order to
sink the instrument to the point D in a liquid of
specific gravity s', and w'' the weight for a liquid
of specific gravity s'', the weights of the liquids displaced
are respectively

$$w + w' \text{ and } w + w''.$$

Therefore, the volumes displaced being the same,

$$s' : s'' :: w + w' : w + w''.$$

(2) *To compare the specific gravities of a solid and a
liquid.*

Let w be the weight which, placed in A, causes the
instrument to sink to D in the liquid.

Place the solid in A, and let w' be the weight, placed in
A, which sinks the instrument to D.

Then place the solid in C, and let the weight w'', placed
in A, sink the instrument to D.

Hence   weight of solid $= w - w'$,

and its weight in the liquid $= w - w''$.

Hence the weight lost, which is the weight of the liquid
displaced by the solid, $= w'' - w'$, and

∴. spec. gravity of solid : that of liquid $:: w - w' : w'' - w'$.

If we take account of the air, we must, as before, add to $w - w'$ the
weight of the air displaced by the solid.

Hare's Hydrometer.

133. This instrument is an application of the principle of the barometer; it consists of two vertical glass tubes leading out of a hollow vessel A, which can be connected with an air-pump.

B and C are two cups in which the lower ends of the tubes are immersed, and which contain the two fluids to be compared.

Let the air in A be partially withdrawn, so that its pressure is diminished from Π the atmospheric pressure to Π'.

Then if D, E be the surfaces of the liquids in the tubes, and F, G in the cups, the weights of the columns DF and EG are each equal to the difference between the atmospheric pressure and the pressure of the air in A.

Hence, if s, s' be the specific gravities, which are proportional to the intrinsic weights,

$$s \cdot DF = s' \cdot EG$$

$$\therefore s : s' :: EG : DF.$$

There is no absolute necessity for an air-pump, as a partial vacuum may be obtained in several other ways.

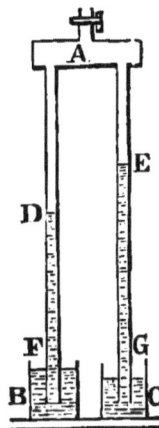

The Stereometer.

134. The name stereometer* has been given to a modified form, by Professor Miller, of Say's instrument for measuring the volumes of small solids.

It consists of two glass tubes, PQ, DB, of equal diameter, cemented into cylindrical cavities communicating with each other at their lower ends in a piece of iron G.

Two apertures lead out of PQ and DB, the one, K, stopped with a screw and the other, L, having a stop-cock.

* Prof. Miller, *Phil. Trans.* Part III. 1856.

The upper end of PQ opens into a cup F, the rim of which is ground plane, so that it can be closed and made air-tight by a well-greased plate of glass. The tube PQ is graduated by lines traced on the glass, and measured downwards from a fixed point P.

The solid to be examined being placed in F, mercury is poured into D, till its surface rises to P, and the cup is then closed by the plate of glass.

The stop-cock L is then opened and the mercury allowed to escape till the difference of the heights of the mercury in the tubes is nearly equal to half the height of the mercury in the barometer. Let M and C mark the height in the tubes; and let u be the volume of the air in F before the solid was placed in it, v the volume of the solid, and h the height of the barometer.

Then we have,

pressure at C : pressure at M :: $h : h - MC$; but these pressures are inversely as the volumes;

$\therefore$ if κ is the section of either tube,

$$u - v + \kappa \cdot PM : u - v :: h : h - MC,$$

and

$$v = u - \frac{h - MC}{MC} \kappa PM.$$

The volume u can be found by a similar process, the cup F being empty, and κ is found by weighing the mercury contained in a given length of the tube.

If the weight w of the solid v be determined, its specific gravity s is given by the relation $w = sv$ (62·5).

135. The screw K is used in the process of finding κ. To do this, the cup is taken off and the tube PQ closed; the tubes are then inverted, the screw K taken out, and mercury is poured in through a slender glass tube inserted in K; this precaution is taken in order to prevent the formation of air-bubbles in PQ.

The end P is then opened and the mercury allowed to run into a glass jar, in which it is weighed.

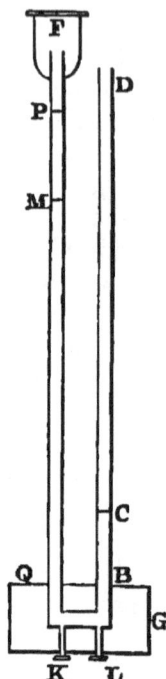

A cubic inch of mercury at 16^0 weighs nearly $3429\frac{1}{2}$ grains, and therefore if w be the weight of a column of mercury a inches in length,

$$w = 3429\tfrac{1}{2} \cdot \kappa a,$$

from which κ is determined in square inches.

Say's instrument consisted of one tube PQ, the lower end being open, so that it could be immersed in a cylindrical vessel of mercury.

The instrument was invented for the purpose of determining the specific gravity of gunpowder : it can be employed in finding the specific gravities of powders or soluble substances, for which the methods which require immersion in water are inapplicable.

EXAMINATION UPON CHAPTER VII.

1. A solid, which is lighter than water, weighs 5 lbs., in vacuo, and when the solid is attached to a piece of metal, the whole weighs 7 lbs. in water ; the weight of the metal in water being 9 lbs., compare the specific gravities of the solid and of water.

2. A solid weighing 25 lbs., in vacuo, weighs 16 lbs. in a liquid A, and 18 lbs. in a liquid B ; compare the specific gravities of A and B.

3. The whole volume of a hydrometer is 5 cubic inches, and its stem is one-eighth of an inch in diameter ; the hydrometer floats in a liquid A with one inch of the stem above the surface, and in a liquid B with two inches above the surface ; compare the specific gravities of A and B.

4. What volume of cork, specific gravity ·24, must be attached to 6 lbs. of iron, specific gravity 7·6, in order that the whole may just float in water ?

5. A body weighs 250 grains in a vacuum, 40 grains in water and 50 grains in spirit ; find the specific gravities of the body and of the spirit.

6. A Sikes's hydrometer floats in water with a given length (a) of its stem not immersed ; it is then placed in a liquid (A), and when a weight w, volume v', is placed on the lower end, it is found that the length of stem not immersed is the same as before ; compare the specific gravity of A with that of water.

7. If a piece of metal weigh in vacuum 200 grains more than in water, and 160 grains more than in spirit, what is the specific gravity of the spirit ?

8. A piece of metal which weighs 15 ounces in water is attached to a piece of wood which weighs 20 ounces in vacuum, and the two together weigh 10 ounces in water, find the specific gravity of the wood.

9. A piece of wood, which weighs 57 lbs. in vacuo, is attached to a bar of silver weighing 42 lbs., and the two together weigh 38 lbs. in water; find the specific gravity of the wood, that of water being 1, and that of silver 10·5.

EXAMPLES.

1. The apparent weight of a sinker, weighed in water, is four times the weight in vacuum of a piece of a material whose specific gravity is required; that of the sinker and the piece together is three times that weight. Shew that the specific gravity of the material is ·5.

2. A hollow cubical metal box, the length of an edge of which is one inch and the thickness one-eighteenth of an inch, will just float in water, when a piece of cork, of which the volume is 4·34 cubic inches and the specific gravity ·5, is attached to the bottom of it. Find the specific gravity of the metal.

3. A crystal of salt weighs 6·3 grains in air; when covered with wax, the specific gravity of which is ·96, the whole weighs 8·22 grains in air and 3·02 in water; find the specific gravity of salt.

4. A Nicholson's hydrometer weighs 6 oz., and it is requisite to place weights of 1 oz. and 1½ oz. in the upper cup to sink the instrument to the same point in two different liquids; compare the specific gravities of the liquids.

5. With the same hydrometer it is found that when a certain solid is placed in the upper cup a weight of 1½ oz. must be placed in the upper cup to sink the instrument in a liquid to a given depth; and that, when the solid is placed in the lower cup, a weight of 3 oz. must be placed in the upper cup to sink the instrument to the same depth; compare the specific gravities of the solid and the liquid, the weight of the solid being 2 oz.

6. A ring consists of gold, a diamond, and two equal rubies, it weighs in vacuo 44¼ grains, and in water 38¾ grains; when one ruby is taken out it weighs 2 grains less in water. Find the weight of the diamond, the specific gravity of gold being 16½, of diamond 3½, of ruby 3.

7. If the price of pure whisky be 16s. per gallon, and its specific gravity be ·75, what should be the price of a mixture of whisky and water, which on gauging is found to be of specific gravity ·8, the specific gravity of water being 1?

8. A common hydrometer has a portion of its bulb chipped off, and, when placed in liquids of densities a, β, γ, it indicates densities a', β', γ' respectively; find the relation between these quantities.

9. A hydrometer marks graduations a, b, c in liquids whose densities are ρ_1, ρ_2, ρ_3 respectively, prove that

$$\frac{b-c}{\rho_1} + \frac{c-a}{\rho_2} + \frac{a-b}{\rho_3} = 0.$$

10. The readings of a common hydrometer when immersed in three different fluids are l_1, l_2, l_3, and the weights which must be placed in the upper cup of a Nicholson's hydrometer in order to sink it to the mark when placed in the different fluids are w_1, w_2, w_3 respectively. Shew that

$$l_1 l_2 (w_1 - w_2) + l_2 l_3 (w_2 - w_3) + l_3 l_1 (w_3 - w_1) = 0.$$

11. Supposing some light material, whose density is ρ, to be weighed by means of weights of density ρ', the density of the atmosphere when the barometer stands at 30 inches being unity; shew that, if the mercury in the barometer fall one inch, the material will appear to be altered by $\dfrac{\rho' - \rho}{(\rho - 1)(30\rho' - 29)}$ of its former weight. Will it appear to weigh more or less?

12. A heavy bottle is filled with a fluid A and weighed in each of two other fluids B, C, the apparent weights being A_b, A_c; it is then filled with the fluid B and weighed in C and A, the apparent weights being B_c, B_a; lastly it is filled with fluid C and weighed in the fluids A and B, the apparent weights being C_a, C_b: shew that

$$A_b + B_c + C_a = A_c + B_a + C_b.$$

CHAPTER VIII.

Mixture of Gases.

136. IF two liquids are mixed together in a vessel, and
if the vessel is left at rest, the two liquids, provided they do
not act chemically on each other, will gradually separate and
finally attain equilibrium with the heavier liquid lowest, and
the lighter liquid superposed upon it.   But if two gases are
placed in communication with each other, even if the heavier
gas be below the other, they will rapidly intermingle until
the proportion of the two gases is the same throughout, and
the greater the difference of density the more rapidly will
the mixture be formed.

Take two different gases, having the same temperature
and pressure, and contained in separate vessels; open a
communication between the vessels, and it will be found
that, unless a chemical action take place, the pressure of the
mixture will be the same as before, provided the temperature
be the same.

We can hence deduce the following proposition :

*If two gases having the same temperature be mixed
together in a vessel of volume* V, *and if the pressures of the
gases when respectively contained in* V, *at the same temperature, be* p *and* p′, *the pressure of the mixture will be* p + p′.

Suppose the gases separate; change the volume of the gas, of which the pressure is p', without change of temperature, until its pressure is p; its volume will then be $\dfrac{p'}{p} V$, by Mariotte's Law.

Now mix the two gases without change of volume, so that the volume of the mixture is $V + \dfrac{p'}{p} V$, or $\dfrac{p + p'}{p} V$; by the preceding experimental fact, the pressure of the mixture will be still p.

Compress the mixture till its volume is V, and when the temperature is the same as before, the pressure, which varies inversely as the volume, will be $p + p'$.

This result is equally true of the mixture of any number of gases.

137. *Two volumes, V, V', of different gases at the respective pressures p, p', are mixed together in a vessel of volume U; it is required to find the pressure.*

Change the volume of each gas to U; their pressures will be respectively

$$\frac{V}{U} p, \quad \frac{V'}{U} p',$$

and therefore the pressure (ϖ) of the mixture will be

$$\frac{V}{U} p + \frac{V'}{U} p'.$$

Hence $\qquad \varpi U = pV + p'V'.$

If the absolute temperatures of the gases before mixture are T and T', and if τ is the absolute temperature, and U the volume after mixture, the pressures of the gases will be respectively

$$\frac{pV}{T} \cdot \frac{\tau}{U}, \quad \frac{p'V'}{T'} \cdot \frac{\tau}{U}.$$

Hence ϖ, the pressure of the mixture, is the sum of these quantities, and therefore

$$\frac{\varpi U}{\tau} = \frac{pV}{T} + \frac{p'V'}{T'}.$$

In the case of the mixture of any number of gases, we have

$$\frac{\varpi U}{\tau} = \Sigma \frac{pV}{T} \ .$$

In Art. (136) we have assumed that Mariotte's law is true of a gas formed by the mixture of two gases; this can be shewn by direct experiment, but is in fact already proved in one case, by the original experiment with atmospheric air, which is itself composed of several different gases. Moreover, the results of the two preceding propositions are borne out by facts.

Vapours.

138. The term vapour is applied to those gaseous bodies, such as steam, which can be liquefied at ordinary pressures and temperatures. There is no difference between the mechanical qualities, as distinguished from the chemical qualities, of vapours and gases, the laws already stated of gases being equally true of vapours within certain ranges of temperature. In fact, there is every reason to believe that all gases are the vapours of certain liquids, but those which are looked upon as permanent gases require the application of extreme cold and of very great pressure to reduce them to a liquid form.

Professor Faraday found that carbonic acid, at the temperature -11°, was liquefied by a pressure of 20 atmospheres*, but that, at the temperature 0°, a pressure of 36 atmospheres was required to produce condensation.

In 1877, M. Pictet succeeded in liquefying oxygen by subjecting it to a pressure of 300 atmospheres, and, at the end of the same year, M. Cailletet effected the liquefaction of nitrogen, atmospheric air, and hydrogen.

139. *Formation of vapour.* If water be introduced into a space containing dry air, vapour is immediately formed, and if the quantity of water be small, and the temperature

---

* An atmosphere denotes the pressure due to a column of mercury 29·9 inches in height.

high, the whole of the water will be rapidly converted into vapour, and in all cases the pressure of the air will be increased by the pressure due to the vapour thus formed.

An increase of temperature, or an enlargement of the space, increases the amount of vapour as long as the supply of water remains; but if the water be removed, an increase of temperature changes the pressure of the vapour in accordance with the general law which regulates the connection between pressure and temperature.

The formation of vapour does not in any way depend upon the presence of air or upon its density, the only effect which the air produces being a retardation of the time in which the vapour is formed. If water be introduced into a vacuum, it is instantaneously filled with vapour, but the quantity of vapour is the same as if the space had been originally filled with air.

Saturation. As long as the supply of water remains as a source from which vapour can be produced, any given space will be always *saturated* with vapour, that is, will contain the maximum quantity of vapour for any temperature; but if the temperature be lowered, a portion of the vapour will be immediately condensed, and become visible in the form of liquid.

The quantity of vapour by which any given space is saturated is proportional to the space for any given temperature; it follows that the pressure, or elastic force, of the vapour is independent of the space it saturates, and depends only on the temperature. No definite law has been discovered connecting the temperature and the elastic force of vapour, but tables have been formed and empirical formulæ constructed for certain ranges of temperature.

140. The laws of the mixture of gases are equally true of the mixture of vapours with each other, or of vapours with gases, provided no condensation take place; or, if any condensation should take place, provided a proper allowance be made for the loss of pressure incurred.

Thus all atmospheric air contains more or less aqueous vapour, and if p be the pressure of dry air and ϖ of the vapour in the atmosphere at any time, the actual atmospheric pressure is $p + \varpi$.

141. *Having given the pressures of a volume* V *of atmospheric air, and of the vapour it contains, to find the volume of the air without its vapour at the same pressure and temperature.*

Let Π be the atmospheric pressure, and ϖ that of the vapour.

Then $\Pi - \varpi$ is the pressure of the air alone when its volume is V;

Hence its volume at a pressure $\Pi = \dfrac{\Pi - \varpi}{\Pi} V$.

142. *Having given the volume* V *of a dry gas at a given temperature under a pressure* p, *to find its volume under the same pressure, when saturated with vapour.*

Let ϖ be the pressure of the vapour.

Then the gas must be allowed to expand until its pressure is $p - \varpi$, the supply of vapour being kept up. The pressure of the mixture is then p, and the volume will be $\dfrac{p}{p - \varpi} V$.

143. *A gas contained in a closed vessel of volume* V *is in contact with water, and its pressure at the temperature* t *is* p; *it is required to determine its pressure when* V *is changed to* V' *and* t *to* t'.

Let ϖ and ϖ' be the pressures of the vapour at the temperatures t and t' respectively, and p' the required pressure.

Then $p - \varpi$ and $p' - \varpi'$ are the pressures of the gas alone, under the two sets of conditions stated.

Hence, if ρ, ρ' be the densities of the gas,

$$p - \varpi = \kappa\rho\,(1 + at),$$
$$p' - \varpi' = \kappa\rho'\,(1 + at'),$$

also
$$\rho V = \rho' V';$$

$$\therefore \quad \frac{p' - \varpi'}{p - \varpi} = \frac{V}{V'} \cdot \frac{1 + at'}{1 + at},$$

whence p' is determined.

If σ, σ' be the densities of vapour under the two conditions,

$$\frac{\varpi'}{\varpi} = \frac{\sigma'\,(1 + at')}{\sigma\,(1 + at)},$$

and combining the two equations,

$$\frac{p' - \varpi'}{p - \varpi} \cdot \frac{\varpi}{\varpi'} = \frac{V\sigma}{V'\sigma'},$$

or
$$\frac{V'\sigma'}{V\sigma} = \frac{p\varpi' - \varpi\varpi'}{p'\varpi - \varpi\varpi'}.$$

If $p\varpi' > p'\varpi$, $V'\sigma'$ will exceed $V\sigma$; i.e. more vapour will have been absorbed by the gas, but if $p\varpi' < p'\varpi$, then $V'\sigma'$ will be less than $V\sigma$, and the gas must therefore, in changing its volume and temperature, have lost a portion of its vapour.

Radiation, Conduction, and Convection of Heat.

144. *Radiation.* All bodies give off heat from their surfaces by what is called radiation, and receive heat by radiation from other bodies. If two bodies at different temperatures are placed near each other, it is an experimental fact that the temperature of one will rise, and of the other diminish until they are both the same.

In a similar manner, if a body is placed in a confined space, the temperature of the body and of the boundary of the space will gradually approximate, the one increasing and the other decreasing till they are the same.

Difference of radiating power. Some bodies radiate heat more freely than others, and the difference appears to depend in great measure on the nature of the surfaces. Thus the leaves of trees and woollen substances radiate heat freely and rapidly, while the radiation from a polished metal surface is very slight.

Generally if the reflecting power of a surface be increased its radiating power is diminished.

145. *Conduction and convection.* There are two other modes of transference of heat from one body to another. Conduction is the term applied to the transference of heat by contact, heat being transmitted through the successive particles of a body, or from one body to another in contact with it. Convection is the actual transference of heat by the motion of fluids or other bodies from one position to another; the heat thus conveyed away from one body may be imparted by contact or radiation from the conveying body to any other.

Thus the handle of a poker, inserted in the fire, is heated by conduction, and in the process of warming rooms by hot air or hot-water pipes the heat is obtained by convection.

There are great differences in the conducting powers of different bodies; liquids generally are weak conductors, but metallic substances have large conducting powers.

The cold felt in placing the hand on a marble mantelpiece is an instance of conduction, the heat being transferred from the hand to the marble.

Woollen substances, glass, and wood, conduct heat very slowly, and this fact is practically taken advantage of in many ways. A heated body rolled up in a woollen cloth may be kept hot for a long time, and ice in a wooden pail, wrapped round with a cloth, will melt very slowly, even in a warm room.

Another instance of a body with very small conducting power is sand; heat is transferred through it so slowly that red-hot shot can be safely carried about in wooden barrows filled with sand.

One of the many useful applications of the non-conducting powers of certain substances is in the construction of *Fire-proof Safes;* a safe of this kind is simply an iron box enclosed within another somewhat larger, the space between being filled up with some non-conducting substance.

146. The explanations above given of the saturating density of vapour, and of the radiation of heat, will enable us to account for many of the ordinary meteorological phenomena, such as the formation of dew, and the fall of rain and snow.

Formation of Dew. Any portion of atmospheric air contains vapour in a greater or less degree, and may be saturated with it; if so, the slightest fall of temperature will produce condensation. If any solid in contact with the atmosphere be cooled down until its temperature is below that which corresponds to the saturation of the air around it, condensation will take place, and the condensed vapour will be deposited in the form of *dew* upon the surface of the body.

This accounts for the *dew* with which the ground is covered after a clear night.

Heat radiates from the ground, and from the bodies upon it, and unless there are clouds from which the heat would be radiated back, the surfaces are cooled and the vapour in the stratum of the atmosphere immediately above condenses and falls in small drops of water on the surface. Any kind of covering will more or less prevent the formation of dew beneath; very little dew, for instance, will be found under the shade of large trees. It will be seen moreover that good radiators are most abundantly covered with dew, very smooth surfaces being almost entirely free from it. This is in accordance with the facts stated above of the radiation of heat.

Hoar-Frost. If after the deposition of dew the temperature fall below the freezing-point, the dew is then frozen and becomes hoar-frost.

The fogs seen at night on low lying or marshy lands are due to the same cause. The air is charged with moisture to saturation, and the cooling of the surface extends sometimes through three or four feet of the atmosphere, producing a thick fog close to the ground, while the air above is quite clear.

147. *Clouds and Rain.* Clouds are formed by the condensation of the vapour in the upper regions of the atmosphere. The reduction of temperature requisite for condensation may occur from several different causes; a mass of air and vapour in motion may rise into a colder region or may come into contact with a larger mass of colder air, so that when the two are mingled together the temperature may not be sufficient to maintain the elasticity of the vapour.

The fact that the clouds remain suspended may be explained in various ways. It seems highly probable that in the process of condensation the vapour assumes the form of small vesicles of water containing air, and therefore not necessarily of greater specific gravity than the medium in which they are formed. Or, again, if the particles do descend, they may, as they fall into a space in which the temperature is higher, be gradually absorbed, and if new vapour be formed above, the appearance of a stationary cloud would consist with the fact of a continuous fall in the constituent particles of the cloud itself.

The cloud which is often seen about the top of a mountain is not unfrequently of this kind. A mass of warm air charged with moisture travels past a mountain, and by contact with it condensation is caused in that portion which is near to the mountain. As the condensed vapour is drifted away, it is again absorbed by the warm air around it, and thus the apparently fixed cloud merely represents a state through which the warm air passes, and from which it emerges.

If a cloud be very highly charged with moisture, and a further reduction of temperature take place, the vapour condenses still further into small drops, and descends in the form of *rain*.

148.   When vapour is being condensed, if the temperature fall below the freezing-point, *snow* is formed; and if rain as it falls pass through a region of the air in which the temperature is below the freezing-point, the drops of rain are congealed and descend in the form of *hail*.

Fogs and *mists* are clouds formed near the earth's surface and in contact with it.   The light summer rain which sometimes falls about sunrise or sunset without the appearance of a cloud is due to the same cause, the air becoming suddenly colder, and the vapour in consequence being rapidly condensed.

149.   *Illustration.*   The phenomena of dew and hoarfrost may be obtained on a small scale by simply putting ice into a glass of water.   The outside of the glass will soon be covered with a delicate dew, which after a short time freezes, and the glass is then covered with hoar-frost.

The explanations of the preceding articles will enable an observer to account for most of the phenomena which depend on the existence of aqueous vapour in the atmosphere.

150.   *Sea and land breezes.*   Winds are partly due to changes of temperature; if, for instance, the air in the neighbourhood of any particular region become heated, it will expand and rise, its place being filled by air from other regions, and hence a wind towards the heated region.

In hot countries on the sea-coast it is noticed that during the day the wind in general blows from the sea, and during the night from the land.   During the day the land becomes heated and retains heat; hence the air above it rises, and the cooler air flows in from the sea.   But during the night the land cools by radiation while the temperature of the sea remains nearly the same; hence the land breeze.

Dew-Point and Hygrometers.

151. *The dew-point* is the temperature at which the vapour in the atmosphere begins to condense.

To determine the dew-point a glass vessel must be cooled until dew begins to be deposited upon it, and its temperature must be then observed; again, observe the temperature at which the dew disappears; a mean between the two may be taken as the dew-point.

152. *Tension of vapour in the air.* The phrase *tension of vapour* is frequently employed to represent the pressure of the vapour. If the dew-point be ascertained we can infer the tension of the vapour in the air by means of the tables before referred to of the relation between the temperature and the saturating density.

For if t' be the dew-point, and ϖ' the corresponding pressure, t the temperature of the air, and ϖ the required pressure

$$\varpi : \varpi' :: 1 + at : 1 + at',$$

and, the pressure being known, the quantity of vapour in the atmosphere can be determined.

153. *Hygrometers* are instruments for determining the quantity of vapour in the atmosphere, or, in other words, the degree of saturation.

This is measured by the ratio of the tension of the vapour in the air to the saturating tension.

Thus if, in the case of Art. (137), ϖ'' be the saturating tension at the temperature t, $\dfrac{\varpi}{\varpi''}$ is the measure required.

Hygrometers may be constructed of any substance which is affected by the amount of moisture in the air, such as a piece of cord which elongates as the quantity of vapour in the air diminishes, or a piece of seaweed, which is exceedingly sensitive to hygrometric changes in the atmosphere.

One of the hygrometers most in use is the *wet and dry bulb Thermometer.* It consists of two mercurial thermometers near each other, one of which is covered with muslin, and kept constantly wet by letting a portion of the muslin drop in a cup of water. The moisture from the muslin evaporates, and, as evaporation is always accompanied by cooling, the wet bulb thermometer falls, and, the drier the air is, the greater will be the difference between the two thermometers. Empirical formulæ and tables have been constructed by means of which the tension of the vapour can be inferred from the readings of the thermometers*.

Dilatation of Liquids.

154. In general, all solid and liquid bodies expand under the action of heat, and contract when heat is withdrawn. We have before had occasion to take account of the expansion of mercury, which is within certain limits proportional to the increase of temperature. This is also the case with solid bodies, such as glass and steel.

For water and aqueous liquids generally, the rate of expansion is not constant for a constant increase of temperature, but beyond a certain limit becomes more rapid as the temperature rises.

Maximum density of water. It is a remarkable property of water that its density is a maximum at a temperature of about 4° C. or 40° F., and whether the temperature increases or decreases from this point, the water expands in volume†.

155. *Freezing.* When the temperature descends to the freezing-point, a still further expansion takes place at the moment of congelation. This is sufficiently proved by the fact that ice floats in water, but it may also be rendered very distinctly evident by a direct experiment. Fill a small iron shell with water, and close the aperture with a wooden plug; if the shell be then exposed to a freezing temperature, the water within will freeze, and at the instant of congelation, the plug will be shot out with considerable violence‡.

Effect of Pressure. It is a remarkable fact that the freezing-point of water is lowered by increase of pressure.

---

* See Mr Glaisher's pamphlet *On the Wet and Dry Bulb Thermometer.*

† The results of Playfair and Joule give 3°·945 C. as the temperature at which the density is a maximum. Prof. Miller, *Phil. Transactions*, 1856.

‡ The temperatures at which liquids freeze are different for different liquids, but fixed for each liquid. Thus mercury freezes at a temperature – 40° C.

This was predicted, from purely theoretical considerations, by Professor James Thomson in 1849, and afterwards established by direct experiment.

156. *Formation of ice on the surface of a lake.* It is known that ice is formed much more rapidly on the surface of shallow than on the surface of deep water; and this fact we can now account for. As the air cools, the water at the surface cools, and being contracted becomes heavier than the water beneath. The surface strata then descend, and the water from beneath rises and becomes cooled in its turn, and this process will go on until the whole of the water has attained its maximum density, after which it will remain stationary, and the upper strata being further cooled will expand and finally congeal. It is clear that the deeper the water is the longer will be the time which elapses before the whole of the water has attained its maximum density.

157. *Ebullition.* When heat is applied to water, it expands gradually until, at a certain temperature, bubbles are formed and steam is given off.

This temperature is the boiling-point, and it has been mentioned before that it depends upon the atmospheric pressure.

The bubbles are first formed by the expansion of the air which water contains. If water be heated from below, the lower strata expand and rise, the upper strata descending and becoming heated in succession, and air-bubbles ascend. As the temperature increases, small bubbles of vapour ascend, but do not always reach the surface, as they may be condensed in the less heated strata above. Finally, larger bubbles are formed, and, the whole mass being heated, ascend to the surface and give off steam, which becomes visible by a slight condensation in the air above.

These bubbles are formed when the tension of their vapour is equal to the pressure they sustain, and this explains why a diminution of atmospheric pressure permits the process of ebullition at a lower temperature; and, on the other hand, that an increase of atmospheric pressure raises the temperature of ebullition.

For instance, under a pressure of two atmospheres, the boiling-point is raised 20° C., and, if the atmospheric pressure be diminished one-half, the boiling-point is lowered about 18°.

This accounts for the fact that water boils at a low temperature on the tops of mountains, and on high table-lands.

Specific Heat.

158. It is found that a certain quantity of heat must be expended in order to raise the temperature of a mass of any substance by a given amount. The requisite quantity of heat depends on the nature of the substance and also on its mass, and for any particular substance it may be at once assumed that the quantity of heat required to raise the temperature one degree is directly proportional to the mass of the substance.

In general, the amount of heat required to change the temperature of a given mass from $t°$ to $(t+1)°$ is the same for all values of t.

Hence for the same substance the quantity of heat expended in changing the temperature from $t°$ to $t'°$

$\propto t' - t$ when the mass is given,

and $\propto$ the mass when $t' - t$ is given,

and therefore generally $\propto m (t' - t)$, if m be the mass.

If this be taken equal to $cm(t' - t)$, c is called the specific heat of the substance, and it is the measure of the amount of heat which will raise by 1° the temperature of the unit of mass.

If two masses m, m', of the same substance, at temperatures t, t', be mixed together, and if τ be the temperature of the mixture, then, since the amount of heat lost by one is gained by the other,

$$m (t - \tau) = m' (\tau - t'),$$
$$\text{or } mt + m't' = (m + m') \tau.$$

159. For different substances the quantity c has different values; thus it is found that water requires about 28 times as much heat as mercury in order to change the

temperature by a given amount; and the specific heat of mercury is therefore less than that of water in the ratio of 1 : 28.

The specific heat of a gas must be considered from two different points of view, for we may suppose the volume of a gas constant, and investigate the amount of heat required to raise the temperature 1°, or we may suppose the pressure constant, the latter supposition permitting the expansion of the gas.

The specific heat in the second case exceeds the specific heat in the first case by the amount of heat disengaged when the gas is suddenly compressed into its original volume.

The specific heat of water is usually taken as the unit, and one of the methods of finding the specific heat of a substance is by immersing it in a given weight of water, and observing the temperature attained by the two substances.

Thus, if M be the mass of a body, T its temperature, and C its specific heat,

m' and m the masses of a vessel and of the water in it, and t their common temperature,

τ the temperature of the whole after immersion, and C' the specific heat of the vessel,

$$CM(T-\tau)=m(\tau-t)+C'm'(\tau-t),$$

since the quantity of heat lost by the body is equal to that gained by the water and the vessel.

If C' be known, this equation determines C; and C', if unknown, can be found by pouring water of a known temperature into the vessel at some other known temperature.

In general, if any number of substances be in thermal contact, and if no heat is lost, the ultimate temperature is given by the equation

$$\tau\Sigma(mc)=\Sigma(tmc).$$

160. *Latent Heat.* It is found that in order to change the state of a body, without changing its temperature, a certain amount of heat must be expended. For instance, in order to convert ice into water, heat must be applied to the ice, and the latent heat of ice is measured by the amount of heat required to convert into water, without change of temperature, one unit of mass of the ice.

1. A cubic foot of air having a pressure of 15 lbs. on a square inch is mixed with a cubic inch of compressed air, having a pressure of 60 lbs. on a square inch; find the pressure of the mixture, when its volume is 1729 cubic inches.

2. State the conditions under which a space is saturated with vapour.

3. A vessel of water is left in a close room for some time; what would be the effect of bringing a quantity of ice into the room?

4. Explain the radiation, conduction, and convection of heat. Why is a cloudy sky not favourable to the deposition of dew?

5. How do you account for the long trail of condensed steam which often follows a locomotive in rainy weather?

6. Explain why it is difficult to heat water from its upper surface.

7. If a piece of ice be put into a glass of water, the external surface is soon covered with a fine dew; account for this fact.

8. Three gallons of water at 45° are mixed with six gallons at 90°; what is the temperature of the mixture?

9. At great altitudes it is sometimes found that a sensation of discomfort is felt; the lips crack and the skin of the hands is roughened; how do you account for these facts?
Can you give any reason why an east wind in England sometimes produces similar effects?

10. Two volumes V, V' of different gases, at pressures p, p', and temperature t are mixed together; the volume of the mixture is U, and its temperature t', determine the pressure.

11. Two vessels contain air having the same temperature t, but different pressures p, p'; the temperature of each being increased by the same quantity, find which has its pressure most increased.
If the vessels be of the same size, and be allowed to communicate with each other, find the pressure of the mixture at a temperature zero.

EXAMPLES.

1. A glass vessel weighing 1 lb. contains 5 oz. of water at 20°, and 2 oz. of iron at 100° is immersed; what is the temperature of the whole, taking ·2 as the specific heat of glass and ·12 of iron?

2. An ounce of iron at 120°, and 2 oz. of zinc at 90°, are thrown into 6 oz. of water at 10° contained in a glass vessel weighing 10 oz.; what is the final temperature, taking ·1 and ·12 as the specific heats of zinc and iron?

3. The pressure of a quantity of air, saturated with vapour, is observed; the mixture is then compressed into half its former volume, and, after the temperature has been lowered until it becomes the same as at first, the pressure is again observed; hence find what would be the pressure of the air (occupying its original space) if it were deprived of its vapour without having its temperature changed.

4. It is related of a place in Norway that a window of a ball-room being suddenly thrown open, a shower of snow immediately fell over the whole of the room. Account for this phenomenon.

5. A drop of water is introduced into the tube of a common barometer which just does not evaporate at the higher of the temperatures t_1°, t_2°.

Given that the elasticity of vapour increases geometrically as the temperature increases arithmetically, shew that if E_1, E_2 be the errors of the above barometer at temperatures t_1°, t_2°, the common ratio of the geometric progression for an increase of temperature of 1° in the case of vapour of water is

$$\left\{\frac{E_1(1-et_1)}{E_2(1-et_2)}\right\}^{\frac{1}{t_1-t_2}};$$

e being the coefficient of expansion for mercury.

6. A closed cylinder contains a piston moveable by means of a rod passing through an air-tight collar at the top of the cylinder. The piston is held at a distance from the bottom of the cylinder equal to one-third of its height, and vapour is introduced above and below of a known pressure, the temperature of the cylinder being such as will support vapour of twice the density without condensation. The piston on being left to itself sinks through two-ninths of the height of the cylinder. Prove that the weight of the piston is five-fourths of the pressure of the vapour upon either side at first.

7. A flask is partially filled with water which is caused to boil until the air is expelled, and then the flask is corked and allowed for a short time to cool. The flask is then placed in cold water, and it is found that the water in it recommences boiling. Explain this phenomenon.

8. A mass of ice at 0° C. is subjected to a pressure of 40 atmospheres, without being allowed to give out or receive heat. Given that the specific heat of ice at constant pressure is about half that of water, that the latent heat of melting is 79, and that the freezing point is lowered ·0075 of a degree for every atmosphere of pressure, shew that rather less than $\frac{1}{300}$th of the mass will be melted.

9. An ounce Av. of silver, specific heat ·06, at 40° F. is immersed in 10 ounces of water at 100° F. Find the greatest amount of heat that the silver can take up from the water; and shew that, if it were all utilised in work, it could lift the silver about 921 yards.

10. A gas saturated with vapour is at a pressure Π. It is then compressed without change of temperature to $\frac{1}{n}$ th of its former volume, and the pressure is then observed to be equal to Π'. Shew that the pressure of the vapour

$$= \frac{n\Pi - \Pi'}{n - 1},$$

and that the pressure of the air in the original volume without its vapour

$$= \frac{\Pi' - \Pi}{n - 1}.$$

11. A vertical cylinder is closed by an air-tight piston, and when the piston is at the top of the cylinder it is filled with vapour at a given pressure : if the temperature be such as would maintain vapour of three times the density, find the least weight of the piston which will not condense any of the vapour.

12. A quantity of ice at 30° F. thaws in the midst of a quantity of air at 60° and reduces the temperature of the air 1° before the water begins to evaporate. Taking the specific gravities of ice, water, and air to be ·96, 1, ·0013, and their specific heats ·5, 1, ·2375, and the latent heat of liquefaction to be 144, find the ratio of the volumes of the ice and the air.

CHAPTER IX.

161. If a cylindrical vessel contain liquid, the pressure of the liquid will produce a strain or tension in the substance of which the vessel is formed. We may imagine the vessel formed of some thin flexible substance, such as silk or paper, and it is obvious that if this substance be not strong enough, it will be torn asunder by the pressure of the liquid.

We proceed to investigate the relation between the pressure and the tension produced by it.

Measure of tension. Imagine a hollow cylindrical vessel formed of a thin flexible substance to be filled with a gas at a given pressure, so that the tension may be the same throughout.

Divide the surface along a generating line, length l, and let T be the whole force required to keep the two parts together;

then, if $T = tl$, t is the tension along any unit of length.

If the cylinder be vertical and filled with water, so that the pressure and therefore the tension vary at different depths, then the tension t at any point, i.e. the rate of tension per unit length, is the limiting value of $\dfrac{T}{x}$, where T is the tension across a length x of the generating line containing the point, when x and therefore T are indefinitely diminished.

162. *A vessel in the form of a circular cylinder with its axis vertical contains fluid; to find the relation between the pressure and tension.*

The pressure being the same at all points of the same horizontal plane, it follows that the tension will be the same at all points of the same horizontal section.

Let PQ, $P'Q'$ be small portions of two horizontal sections very near each other, PP' and QQ' being vertical. The dimensions of PQ' are taken so small·that the pressure and tension at all points of it are sensibly the same.

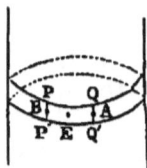

Let p, t be the pressure and tension; then $t.PP'$, $t.QQ'$ are the horizontal forces acting on the portion PQ' of the surface at the middle points A, B of its ends, and these forces must counterbalance the pressure of the liquid, which is $p.PP'.PQ$.

This resultant pressure acts in the direction CE bisecting the angle ACB, and the two tensions in the directions of the tangents at P and Q.

Hence, resolving the forces in the direction CE,

$$p.PP'.PQ = 2t.PP' \sin \tfrac{1}{2} ACB$$

$$= 2t.PP' . \frac{1}{2} . \frac{AB}{r} \quad = \frac{t}{r} . PP'.PQ,$$

if r be the radius of the cylinder,

and $\qquad\qquad\qquad \therefore t = pr.$

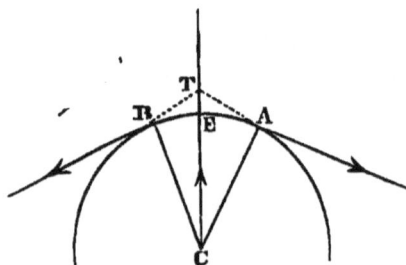

If the cylinder contain a gaseous fluid of which the pressure is sensibly the same throughout its mass, the relation $t = pr$ is true at every point, whether the axis be vertical or not.

This result can also be obtained by considering the equilibrium of a semi-circular portion of thickness PP', for the resultant pressure will be parallel to the tensions at the two ends, and will be equal to the pressure on the projected area $2r.PP'$, so that

$$2t.PP' = p.2r.PP', \text{ or } t = pr.$$

163. If the pressure is different at different points of the arc which is the cross section of a cylindrical vessel, the circular cylindrical form is not a form of equilibrium.

But, if we take r to be the radius of curvature at any point E at which the pressure is p, it can be shewn, exactly as in the previous article, that

$$t = pr.$$

Taking any cross section, the tension will be the same at all points of this section, because the fluid pressure is normal to the surface.

Hence, knowing the tension and the law of pressure, the curvature at every point of the cross section is determined, and the shape of the curve can be found.

For example consider *the Lintearia*, which is the form assumed by a rectangular piece of a thin membrane, two opposite sides of which are fastened to the sides of a box, while the other sides fit the box closely, so that liquid can be poured in without escaping.

The figure is a section of the cylindrical surface so formed, by a plane perpendicular to its generating lines, BC being the surface of the liquid.

The tension (t) along BAC is constant, because the liquid pressure is normal, and if r be the radius of curvature at P,

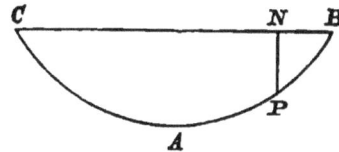

$$t = pr = wr.PN, \quad \therefore \ \frac{1}{r} \propto PN,$$

i.e. the curvature at P is proportional to the depth below the surface.

This curve is the same as the *Elastica*, the curve formed by a bent rod, and is also, as will be seen subsequently, the same as the *Capillary curve*.

164. *A spherical surface contains gas at a given pressure, it is required to find the tension at any point.*

From symmetry we may take the tension to be the same at every point.

Moreover, if any line be drawn on the surface we may assume that the tension between the two portions parted by that line acts in a direction perpendicular to it.

Consider the equilibrium of a hemisphere, under the action of the tension $2\pi rt$, and of the resultant pressure, which is equal to the pressure on a circular area of radius r; we then have

$$2\pi rt = \pi r^2 p, \text{ or } 2t = pr.$$

Hence it appears that a spherical vessel is twice as strong as a cylindrical vessel of the same material and the same radius.

165.   We have not compared with each other the tensions of vessels formed of substances of different thickness. To do this it will be seen that for a given value of the tension t, as we have measured it, the intrinsic stress of any substance will be diminished by increasing the thickness.

Now if e be the thickness of any flexible lamina, and if $t = e\tau$, then τ will be the tension of an unit of area of the section, and for the comparison of different thicknesses, this latter measure of tension must be employed.

Ex.   *A bar of metal one square inch in section can sustain a weight of 1000 lbs., and of this metal a cylinder is made one-twentieth of an inch in thickness, and one foot in diameter; find the greatest fluid pressure which the cylinder can sustain.*

In this case $e = \frac{1}{20}$ and $r = 6$;

also the greatest possible value of τ is 1000 :

$$\therefore p = \frac{e\tau}{r} = 8\frac{1}{3} \text{ lbs. wt.}$$

Hence a force per square inch equal to the weight of $8\frac{1}{3}$ lbs. is the greatest pressure which can be applied without bursting the cylinder.

166.   *A conical vessel, formed of a flexible substance, is held by the rim with its vertex downwards, and is filled with liquid; it is required to find the tension at any point in the direction of the generating line passing through the point.*

Let PP' be a horizontal section of the cone. It is obvious that along the section PP' the tension is the same at any point and is in the direction of the generating line through that point.

Let then t be the tension, which is at all points of the circle PP' in a direction inclined at an angle a to the vertical, if $2a$ be the vertical angle.

The vertical resultant of the tension on the whole circle PP', that is, $2\pi . PN . t \cos a$, is equal to the resultant vertical pressure on the surface POP'.

Now this pressure

$$= \text{weight of fluid } POP' + \text{weight of fluid } P'Q$$

$$= w\left(\frac{1}{3}\pi PN^2 . ON + \pi PN^2 . PQ\right),$$

and therefore if $ON = x$, and $OE = h$,

$$2\pi x \tan a . t \cos a = w\pi x^2 \tan^2 a \left(\frac{x}{3} + h - x\right),$$

or

$$t = \frac{1}{2} w \frac{\sin a}{\cos^2 a}\left(hx - \frac{2x^2}{2}\right).$$

Since

$$hx - \frac{2x^2}{3} = \frac{2}{3}\left\{\frac{9h^2}{16} - \left(x - \frac{3h}{4}\right)^2\right\}$$

it follows that t has a maximum value when $x = \frac{3h}{4}$.

A little consideration will shew that there is a horizontal tension at all points along a generating line, in a direction perpendicular to that line, but the investigation of this other tension would be beyond the limits which must be assigned to an elementary course, and must therefore be deferred to treatises taking a higher range.

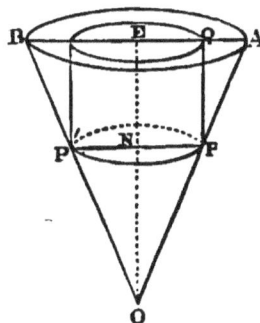

EXAMPLES.

1. Two vertical cylinders of the same thickness and the same material, contain equal quantities of water; compare their greatest tensions.

2. Two cylindrical boilers are constructed of the same material, the diameter of one being three times that of the other, and the thickness of the larger one twice that of the other; compare the strengths of the boilers.

3. A bar of metal, one-fourth of a square inch in section, can support a weight of 1000 lbs.; find the greatest fluid pressure which a cylindrical pipe made of this metal can sustain, the diameter being 10 inches and the thickness one-tenth of an inch.

4. Equal quantities of the same material are formed into two thin spherical vessels of given radii; compare the greatest fluid pressures they will sustain.

5. The natural radius of an elastic spherical envelope containing air at atmospheric pressure is a, and, when a certain quantity of air is forced into it, its radius is b. It is then placed under an exhausted receiver and its radius becomes c. Find the quantity of air forced in, supposing that the increase of tension of the envelope varies directly as the increase of its surface.

6. The top of a rectangular box is closed by an uniform elastic band, fastened at two opposite sides, and fitting closely to the other sides; the air being gradually removed from the box, find the successive forms assumed by the elastic band, and when it just touches the bottom of the box, find the difference between the external and internal atmospheric pressures.

7. A vertical cylinder formed of a flexible and inextensible material contains water; find the tension at any point.

If this flexible cylinder be put into a square box, the width of which is less than the diameter of the cylinder, and water be then poured in to the same height as before, find the change in the tension at any depth.

8. An elastic and flexible cylindrical tube contains ordinary atmospheric air; if the ends be kept closed, and the pressure of the air inside be increased by a given amount, find the increase in the radius of the cylinder.

If the radius be doubled by a given increase of pressure, prove that the modulus of elasticity is in that case twice the tension that would have been produced in the cylinder, if inelastic, by the same increase of pressure.

9. An inelastic flexible cylindrical vessel, closed rigidly at the top, is filled with water, and the whole rotates uniformly about the axis of the cylinder, which is vertical; find the tension at any point.

10. A cast-iron main, 9 inches in diameter internally, is employed for the transmission of water to a reservoir at a height of 300 feet. Find the least thickness of iron which can be employed, subject to the condition that the tension of the metal shall not exceed 5 tons weight per square inch, assuming that a cubic foot of water contains 62·5 lbs.

11. The tensile strength of cast-iron being 16000 lbs. weight per square inch of section, find the thickness of a cast-iron water-pipe whose internal diameter is 12 inches, that the stress upon it may be only one-eighth of its ultimate strength when the head of water is 384 feet.

12. Supposing the cylinders of a Bramah's Press made of the same material, and the stress to be the same in each, what should be the ratio of the thicknesses of the cylinders?

13. A cylindrical vessel is formed of metal a inches thick, and a bar of this metal of which the section is A square inches, will just bear a weight W without breaking. If the cylinder be placed with its axis vertical, find how much fluid can be poured into it without bursting it.

14. An elastic tube of circular bore is placed within a rigid tube of square bore which it exactly fits in its unstretched state, the tubes being of indefinite length; if there be no air between the tubes and air of any pressure be forced into the elastic tube, shew that this pressure is proportional to the ratio of the part of the elastic tube that is in contact with the rigid tube to the part that is curved.

15. A spherical elastic envelope is surrounded by, and full of, air at atmospheric pressure (Π), when an equal amount is forced into it. Prove that the tension at any point of the envelope then becomes

$$\frac{\Pi}{2r'^2}(2r^3 - r'^3),$$

where r, r' denote the initial and final radii.

16. A hemispherical bag of radius c, supported at its rim, is filled with water; prove that, at the depth x, the tension in direction of the meridian section is proportional to

$$(c^2 + cx + x^2)/(c + x).$$

17. A bag, in the form of a paraboloid, formed of thin flexible substance, is supported by its rim, and is filled with water; find the tension at any point in the direction of the tangent to the generating parabola at that point. Hence prove that the tension in every direction at the vertex $= wah$, if h is the depth of the bag, and $4a$ the latus rectum.

18. If the same bag, when filled, be closed and inverted, prove that the tension at any point P, in the direction of the generating parabola, varies as $AN . \sqrt{SP}$, A being the vertex of the bag, S the focus and AN the depth of P below the vertex.

19. An elastic spherical envelope is surrounded by air saturated with vapour. When the air within it is at a pressure of two atmospheres it is found that its radius is twice its natural length, and again the radius is three times its natural length when the envelope contains 77 times as much air as it would if open to the air; assuming that the tension at any point varies as the extension of the surface, prove that one twenty-fifth of the pressure of the air is due to the vapour which it contains.

CHAPTER X.

167. When a glass tube, of very small bore, with its two ends open, is dipped in water it is observed that the water rises in the tube, and that it is in equilibrium with the surface of the water inside at a higher level than the surface outside. If the tube is dipped in mercury, it is found that the mercury inside is in equilibrium at a lower level than the mercury outside.

In either case, the ascent, or depression, is greater if the experiment be made with tubes of smaller bore.

If the surface of water be examined close to the vertical side of a vessel containing it, the surface will be found to be curved upwards, the water appearing to cling to, and hang from the wall, at a definite angle.

Phenomena of this kind, with others, such as those presented by drops of liquid, or by liquid films, are grouped together as being instances of *Capillary Action.*

Consider the equilibrium of a thin column of liquid PQ, as in the figure.

If Π be the atmospheric pressure, the pressure at Q

$$= \Pi - w \cdot QN.$$

Hence, taking κ as the cross section, the column PQ is acted upon by gravity, by the atmospheric pressure $\Pi\kappa$ downwards, and by the pressure $(\Pi - w \cdot QN)\kappa$ upwards.

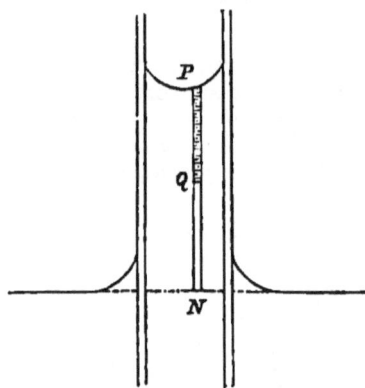

The weight of the column PQ being $w\kappa . PQ . \kappa$, it follows that the resultant of these three forces is $wPN\kappa$ downwards, and this force must in some way be counterbalanced.

This suggests the theory of the existence of a surface tension, the vertical resultant of which, acting on the upper boundary, at P, of the column, will exactly counterbalance the weight of the column PN.

Various facts support the idea of the existence of a surface tension. The familiar experiment of gently placing a needle on the surface of water, on which it will sometimes float, is a case in point. The needle appears to be supported on a thin membrane, which bends beneath its weight.

In summer weather insects may be seen on the surface of water, apparently indenting, without breaking through, the superficial membrane.

As the results of observation and experiment we can state three laws relating to surface tension.

(1) *At the bounding surface separating air from any liquid, or between two liquids, there is a surface tension which is the same at every point and in every direction.*

(2) *At the line of junction of the bounding surface of a gas and a liquid with a solid body, or of the bounding surface of two liquids with a solid body, the surface is inclined to the surface of the solid body at a definite angle, depending upon the nature of the solid and the liquids.*

(3) *The surface tension is independent of the curvature of the surface, but, if the temperature be increased, it diminishes.*

In the case of water in a glass vessel the angle is acute; in the case of mercury in a glass vessel it is obtuse. In the first case the water is said to wet the glass; in the latter the mercury does not wet the glass.

When three fluids are in contact with each other, assuming that they do not mix together, their bounding surfaces will meet in a line, which may be straight or curved.

If we consider a short element of this line, there will be three surface tensions, in planes passing through it, counterbalancing each other; and therefore, if T_1, T_2, T_3 are the surface tensions between the three pairs, and α, β, γ the

angles between their directions, the conditions of equilibrium
are that

$$T_1 : T_2 : T_3 :: \sin \alpha : \sin \beta : \sin \gamma.$$

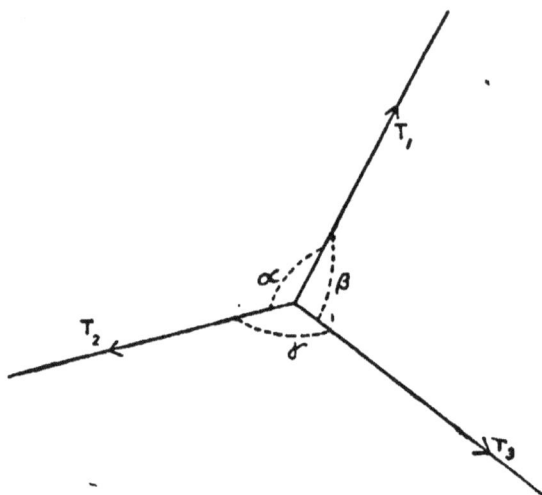

Rise of a liquid between two plates.

168. Take the liquid to be such as to wet the plates, as
in the case of water and glass.

Let the first figure of Art. 167 represent a vertical section
perpendicular to the plates. If T is the surface tension, α
the angle of capillarity, h the mean rise, and d the distance
between the plates, we have, for the equilibrium of one unit
of breadth of the liquid,

$$2T \cos \alpha = whd,$$

so that h varies inversely as d.

Apparent attraction of the two plates to each other.

Since the pressure at any point of the surfaces of the glass
inside, which is above the level of the liquid outside, is less
than the atmospheric pressure, it follows that the resultant
horizontal force on each plate is inwards, and therefore the
plates, if allowed to move, will approach to and cling to each
other.

If the liquid is such as not to wet the plates, as in the case of mercury and glass, the plates will be pressed

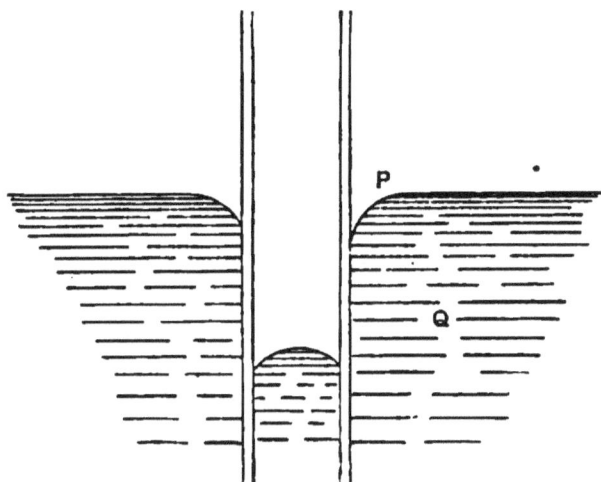

outwards by atmospheric pressure and inwards by pressure greater than the atmospheric pressure.

They will in this case apparently attract each other.

If however the liquid is such that it wets one plate and not the other, the level of the wetted surface E inside will

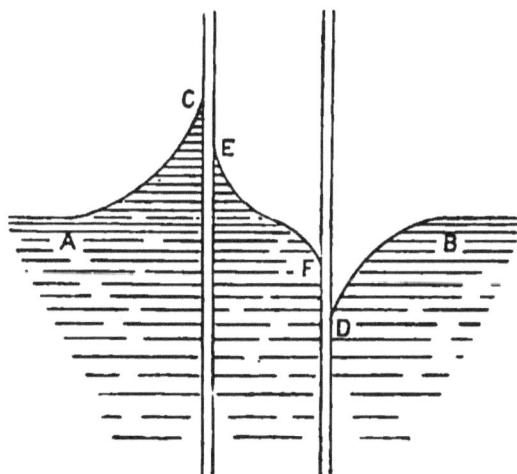

be below C, that of the outside surface, and for the other plate the level F will be above the level D.

The resultant horizontal force on each plate will be outwards and the plates will apparently repel each other.

169. *Rise of a liquid in a circular tube.*

Taking the figure of page 158 as a section through the axis, and r as the radius of the tube, we have

$$2\pi r T \cos a = w\pi r^2 h,$$

and therefore h varies inversely as r.

It will be seen that the rise in a circular tube of radius r is the same as the rise between two plates at a distance r.

In each case the pressure at any point of the suspended column is less than the atmospheric pressure, and, if the column were high enough, this pressure would merge into a state of tension, which would still follow the law of fluid pressure, of being the same, at any point, in every direction.

The rise of sap in trees may perhaps afford an instance of this state of things.

170. *The Capillary Curve* is the form assumed by the liquid near a vertical wall.

Let PN be the height above the level of the water of a point P of this curve, and consider the equilibrium of the column $PQLN$, taking one unit of breadth perpendicular to the plane of the paper.

The resultant of the tensions at P and Q is in direction of the normal at the middle point of PQ, and, if r be the radius of curvature, it is equal to $T \cdot \dfrac{PQ}{r}$.

The vertical component of this resultant being equal to the weight of the column,

$$T \cdot \frac{PQ}{r} \cos \theta = w \, PN \cdot NL,$$

where θ is the inclination of the normal to the vertical;

$$\therefore \text{ since } NL = PQ \cos \theta,$$

$$T = wr \cdot PN,$$

i.e. the curvature at P is proportional to PN.

This is the property which we found to be true of the Lintearia, and which can be shewn to be also the characteristic property of the Elastica.

171. *Needle floating on the surface of water.*

It is well known that a small needle, if placed gently on the surface of still water, will float. The reason is that the surface is slightly indented and that the surface tensions, at the lines of contact of the needle with the surface, have a vertical component.

The resultant of those tensions, combined with the resultant pressure of the liquid, sustains the weight of the needle.

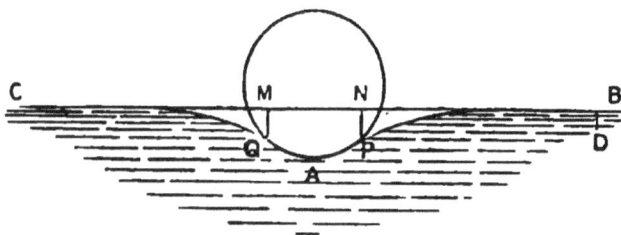

Thus, the figure representing a cross section of the needle and the water surface, the weight of the needle is counteracted by the two surface tensions t, acting in the directions of the tangents at P and Q to the water surface, and the resultant pressure upwards of the liquid, which is equal to the weight of the volume $PAQMN$ of water.

We have the further condition of equilibrium that, if we draw the horizontal line PD through P and the vertical line

BD through B, the horizontal component of the tension at P, together with the horizontal water-pressure on BD, is equal to the surface tension at B.

Let h be the height of the axis of the needle above the level surface of the water, c the radius of the needle, and 2θ the angle POQ.

The area $PAQMN = $ segment $PAQ + $ rectangle $PQMN$;

$\quad$ segment $PAQ = $ sector $POQ - $ triangle POQ

$\quad$ and $PQMN = 2 \{$triangle $POQ\} - h \cdot 2c \sin \theta$;

$\quad \therefore PAQMN = c^2\theta + c^2 \sin \theta \cos \theta - 2ch \sin \theta.$

Hence taking α as the acute angle of capillarity, and W as the weight of the needle, the first condition gives the equation, $2t \sin (\theta - \alpha) + wc (c\theta + c \sin \theta \cos \theta - 2h \sin \theta) = W$, and, from the second condition, we obtain

$$t \cos (\theta - \alpha) + w \frac{1}{2} (c \cos \theta - h)^2 = t,$$

or, $$4t \sin^2 \frac{1}{2} (\theta - \alpha) = w (c \cos \theta - h)^2.$$

It should be noticed that the surface of the needle must be, as it usually is, somewhat oily or greasy, so that its surface is not wetted by water. A highly polished needle will sink at once.

If two needles are floating in water side by side and near each other they will run together, the reason being, as in the case of the two plates in Art. 168, that the liquid between the needles is entirely above the outside level, and therefore there is an excess of horizontal pressure inwards.

Liquid Films.

172. Liquid Films possess the characteristic property that the tension is the same at every point, and in every direction.

It must be carefully noticed that, since a film has two surfaces, the tension of the film is twice the surface tension of the liquid.

Liquid films may be formed, and examined, by shaking a clear glass bottle containing some viscous liquid, or by dipping a wire frame into a solution of soap and water, and slowly drawing it out.

In this way films, apparently plane, can be obtained, shewing that the action of gravity is unimportant in comparison with the tension of the film.

These films give way and break under the least tangential action, and we therefore infer that the tension across any line is normal to that line.

We can hence deduce the property above stated. For, considering a small triangular portion, the actual tensions on the sides must be proportional to the lengths of the sides, and therefore the measures of the three tensions are the same.

If one part of the boundary of a plane film be a light thread, we can prove that it will take the form of an arc of a circle.

Since the tension of the film is at all points normal to the thread, it follows that the tension, t, of the thread is constant.

Let τ be the intrinsic tension of the film, and consider the element PQ of the thread; for equilibrium, if r be the radius of curvature,

$$t \frac{PQ}{r} = \tau \cdot PQ,$$

and therefore r is constant.

173. *Energy of a plane film.*

In drawing out a film a certain amount of work is expended, and this represents the energy of the film.

Consider for instance a plane rectangular film $ABCD$, bounded by wires, and imagine the wire CD moveable on AC and BD;

then releasing CD the film will draw CD towards AB, and the work done, if t be the tension, will be $t \cdot CD \cdot AC$. But, if S be the superficial energy per unit of area, the actual energy is $S \cdot CD \cdot AC$;

$$\therefore S = t,$$

i.e. the superficial energy per unit of area is equal to the tension per unit of length. (Maxwell's *Heat*, Chapter xx.)

174. *Soap-bubbles.*

If t is the tension of the film of a soap-bubble, and p the difference between the internal and external air pressures,

$$2t = pr.$$

Energy of a soap-bubble.

The work done in expanding a soap-bubble from radius r to a radius r' slightly greater is

$$p \cdot 4\pi r^2 (r' - r), \text{ or } 8\pi t r (r' - r).$$

Hence the whole work done in the formation of a bubble of radius $c = \Sigma\, 8\pi t r\, (r' - r)$,

and, taking $r' - r = \dfrac{c}{n}$ and $r = \dfrac{mc}{n}$,

this $= 8\pi t c^2 \sum\limits_{1}^{n} \dfrac{m}{n^2} = 4\pi t c^2$, when n is indefinitely increased,

and therefore the superficial energy $= t$.

Table of Surface Tensions.

In the following list of surface tensions the first column gives the surface tensions in milligrammes weight per millimetre of length, and the second column gives the tensions in grains weight per inch.

It will be observed that one milligramme per millimetre is the same as $\dfrac{1}{64 \cdot 799}$ grains per $\dfrac{1}{25 \cdot 4}$ inch, which is $\cdot 392$ gr. per inch.

The first column is taken from a table by Van der Mensbrugghe, and the second column is obtained from the first by means of the multiplier $\cdot 392$.

Surface Tensions in French and English measures.

Distilled water at 20° C.	7·3	2·86
Sulphuric ether	1·88	·737
Absolute alcohol	2·5	·97
Olive oil	3·5	1·37
Mercury	49·1	18·8
Solution of Marseilles soap,		
1 of soap to 40 of water	2·83	1·11

EXAMPLES.

1. Having given that in a glass tube ·04 inch in diameter the capillary elevation of water is 1·2 inch, and of alcohol ·5 inch, find what it will be for each liquid in a tube ·25 inch in diameter, and in a tube ·05 inch in diameter.

2. In a glass tube ·08 inch in diameter, the capillary depression of mercury is ·15 inch ; find what it will be in a tube ·025 inch in diameter.

3. Two spherical soap-bubbles are blown, one from water, and the other from a mixture of water and alcohol ; if the tensions per linear inch are equal to the weight of $2\frac{1}{2}$ grains and $\frac{7}{12}$ grain respectively, and if the radii are $\frac{7}{8}$ inch and $1\frac{1}{3}$ inch respectively, compare the differences, in each case of the internal and external air pressures.

Also compare the quantities of atmospheric air contained in the bubbles.

4. The superficial tensions of the surfaces separating water and air, water and mercury, mercury and air, are respectively in the ratio of the numbers, 81, 418, 540 ; what will be the effect of placing a drop of water upon a surface of mercury ?

5. Explain why it is that a drop of oil, placed on the surface of water, spreads out rapidly into a layer of extreme tenuity.

6. Prove that if a light thread with its ends tied together forms part of the internal boundary of a plane liquid film the thread will take the form of a circle.

7. If two soap-bubbles, of radii r and r', are blown from the same liquid, and if the two coalesce into a single bubble of radius R, prove that the tension of the bubble is to the atmospheric pressure in the ratio of

$$R^3 - r^3 - r'^3 \text{ to } 2(r^2 + r'^2 - R^2).$$

8. If the pressure inside a soap-bubble is p_0 when its radius is r_0, and if after a volume of air $\frac{4}{3}\pi a^3$ at atmospheric pressure is forced into it the pressure and radius become p and r ; find p and p_0 in terms of a, r_0, r and the atmospheric pressure.

9. If water be introduced between two parallel plates of glass at a very small distance d from each other, prove that the plates are pulled together with a force equal to

$$\frac{2At \cos a}{d} + Bt \sin a,$$

A being the area of the film, and B its periphery.

10. A soap-bubble is filled with a mass m of a gas the pressure of which is $\kappa\rho$, ρ being its density. The radius of the bubble is a when it is first placed in air. The temperature remaining unchanged, prove that, for a slight increase ϖ of the atmospheric pressure, the small decrease x of the radius is given by the equation

$$\frac{\varpi a^2}{x} = \frac{9m\kappa}{4\pi a^2} - 2t,$$

where t is the tension of the film.

11. A small cube of volume a^3 floats, with its upper face horizontal, in a liquid such that its angle of contact with the surface of the cube is obtuse and equal to $\pi - a$.

If w is the intrinsic weight of the liquid and w' of the cube, and if wc^2 is the surface tension, prove that the cube will float if

$$\frac{w'}{w} < 1 + 4\,\frac{c^2}{a^2}\cos a.$$

12. Find the condition that a small cylinder may float in water, the angle of capillarity being obtuse.

13. A cylindrical rod hangs down vertically so as to be partly above and partly below the surface of a liquid resting in a large vessel. Shew that its apparent weight is equal to its weight in air increased by the (positive or negative) quantity by which the weight of the volume of liquid drawn up above the plane level exceeds the weight of a quantity of liquid equal in volume to the portion of the solid below the plane level. Alter the statement to suit cases in which the solid depresses the liquid.

14. Prove that, when liquid rises in a fine capillary tube, the potential energy, which is thereby produced, of the liquid, is independent of the radius of the tube.

15. Two spherical soap-bubbles, made from the same mixture of soap and water, are allowed to form a single soap-bubble; prove that a diminution of surface takes place, and an increase of volume, and that the numerical expressions for the decrease and increase are in a constant ratio to each other.

16. Two soap-bubbles are in contact; if r_1, r_2 be the radii of the outer surfaces, and r the radius of the circle in which the three surfaces intersect, prove that

$$\frac{3}{4r^2} = \frac{1}{r_1^2} + \frac{1}{r_2^2} - \frac{1}{r_1 r_2}.$$

CHAPTER XI.

THE EQUILIBRIUM OF FLUIDS UNDER THE ACTION OF ANY GIVEN FORCES.

175. IN any field of force the measure of the force at any point is the force which would be exerted upon the unit of mass supposed to be concentrated at that point.

As in Art. (10), it can be shewn that the pressure at any point is the same in all directions; for if we consider the equilibrium of a very small prism, the forces at all points of the prism will be ultimately equal and parallel, and the case then becomes the same as that of a prism under the action of gravity.

176. *The measure of the force at a point, in a given direction, multiplied by the density, is equal to the rate of change, per unit of length, of the pressure in that direction.*

If P be the point, take any length PQ in the direction considered and describe a very thin cylinder about PQ.

The equilibrium of this cylinder is maintained by the pressures on its ends and on its curved surface and by the external forces in action.

Therefore the difference of the pressures on the ends P and Q is equal to the force on the cylinder in the direction PQ. Hence, if κ is the cross section, and if PQ is very small, we may consider the density of the cylinder uniform, and we may also take f, the resolved part of the force in the direction PQ, to be the same at all points of PQ, we then obtain, if p and p' are the pressures,

$$(p' - p)\,\kappa = \rho\kappa \,.\, PQ\,.\,f$$

so that

$$\rho f = \frac{p' - p}{PQ},$$

which is the rate of change of pressure.

177. Def. Surfaces of equal pressure are surfaces at all points of which the magnitude of the pressure is the same.

Surfaces of equal pressure are at every point perpendicular to the resulting force.

To prove this, consider two consecutive surfaces of equal pressure containing between them a stratum of fluid, and let a small circle be described about a point P in one surface, and a portion of the fluid cut out by normals to that surface through its circumference.

This small cylinder of fluid is kept at rest by the external force and by the pressures on its ends and on its circumference.

The pressures at all points of the circumference being equal, the pressures on the two ends must be counter-balanced by the external force, which must therefore act in the direction of these pressures, i.e. perpendicular to the surface of equal pressure.

Again, if d be the distance at P between the consecutive surfaces, we have, as before,

$$\rho \kappa d f = (p' - p)\, \kappa,$$

f being the magnitude of the resultant force on unit mass,

so that $\rho d \propto \dfrac{1}{f}$, and in the case of a homogeneous liquid,

$$d \propto \frac{1}{f}.$$

178. If in any field of force a particle be in contact with a smooth surface, it will be in equilibrium if the normal to the surface coincide with the direction of the resultant force.

Surfaces of equilibrium are therefore at all points perpendicular to the resultant force.

If a particle be moved over a surface of equilibrium no

work is done against the force, and these surfaces are therefore surfaces of equal energy, or equipotential surfaces.

If a particle of mass unit be carried along the normal from one surface to another the work done is $f . PQ$, which is the change of energy and is constant;

$$\therefore f . PQ \text{ is constant.}$$

Surfaces of equal pressure are also surfaces of equal density;

For ρfd is constant and we have just shewn that fd is constant, $\therefore \rho$ is constant.

EXAMPLES.

179.   1.   *A mass of liquid at rest under the action of a force to a fixed point varying as the distance from that point.*

The surfaces of equilibrium, and therefore of equal pressure, are clearly concentric spheres, and the free surface is a sphere.

To find the pressure at any point P, take a thin cylindrical column from P to the surface and observe that its equilibrium is maintained by the pressure at the end P counterbalancing the attractive force.

If κ be the cross section, $OP=r$, $OA=a$, and if μr be the force at the distance r,

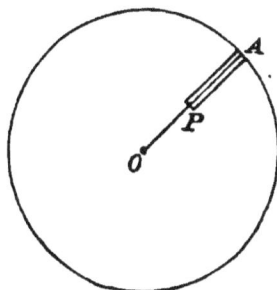

$$p\kappa =\text{force on the column } AP$$

$$=\rho\kappa (a^2 - r) \mu \frac{a+r}{2}, \text{ by Leibnitz's theorem ;}$$

$$\therefore \qquad\qquad p=\tfrac{1}{2}\mu\rho (a^2 - r^2).$$

The Pressure on a diametral plane = Force on a hemisphere

$$=\frac{2}{3} \rho\pi a^3 . \mu . \frac{3a}{8} = \frac{1}{3} \mu\rho\pi a^4.$$

2.   *Liquid at rest under the action of forces to any number of centres varying as the distances.*

It follows from Leibnitz's theorem that the resulting force is directed to a fixed point and varies as the distance from that point; this case is therefore the same as the preceding.

3. *Heavy liquid at rest under the action of a force to a fixed point varying as the distance from that point.*

Taking μc to represent the weight of unit mass, the resultant force on any element of the liquid is directed to a fixed point at a depth c below the centre of force, and the surfaces of equal pressure are therefore spheres having this lower point as centre.

4. *Liquid at rest under the attraction of a straight rod, the molecules of which attract with forces varying inversely as the square of the distance.*

If AB be the rod, it can be shewn by elementary geometry that the direction of the resulting attraction at any point P bisects the angle APB; from this it follows that the surfaces of equal pressure are confocal spheroids, having their foci at A and B.

5. *Liquid at rest under the action of gravity and of forces perpendicular to the horizontal plane base of the vessel containing the liquid and proportional to the distance from that base.*

Let a be the height of the free surface above the base, and consider the equilibrium of a vertical cylinder of liquid extending from the surface to the depth $a-z$.

Then if p is the pressure at the height z above the base and κ the cross-section of the cylinder, the equation of equilibrium is

$$p\kappa = w(a-z)\kappa + \rho\kappa(a-z) . \mu\frac{a+z}{2},$$

by Leibnitz's theorem;

$$\therefore \qquad p = w(a-z) + \tfrac{1}{2}\mu\rho(a^2 - z^2).$$

6. *Heavy homogeneous liquid, every particle of which attracts every other with a force which varies as the distance, fills a sphere; find the surfaces of equal pressure.*

7. *A sphere is filled with fluid at rest but each particle of it is acted on by a force tending to a point on the sphere and varying as the distance. If the pressure at the other end of the diameter through this point be zero, prove that the pressure at the given point is to the pressure at the end of a perpendicular diameter as 2 : 1.*

8. *A mass of liquid is at rest on the outside of a sphere under the action of forces such that the force on any element m of the liquid is directed to the centre of the sphere, and is equal to mf, where f is the same for all the elements of the liquid. Prove that the resultant pressure on a band of the sphere cut off by any two parallel planes is proportional to the mass of a volume of the liquid which would be contained between two coaxal cylinders whose radii are equal to the distances of the planes from the centre of the sphere, and whose heights are each equal to the depth of the given liquid.*

CHAPTER XII.

SOLUTIONS OF VARIOUS PROBLEMS.

180. *Centre of Pressure.* A general expression can be obtained for the depth of the centre of pressure of any plane area.

Let the area be divided by horizontal lines into a number of very small portions, and let a be the area of one of these portions and z its depth below the surface.

Then the pressure upon it $= wza$, and if $\bar{z}$ be the depth of the centre of pressure, we have by the usual formula for the centre of a system of parallel forces,

$$\bar{z} = \frac{\Sigma wza \cdot z}{\Sigma wza} = \frac{\Sigma (z^2 a)}{\Sigma (za)},$$

$w\Sigma (za)$ being the pressure on the whole area.

Ex. An isosceles triangle is immersed vertically, its base being horizontal and its vertex A at a depth c below the surface.

Let
$$AD = h,$$

$$AN = \frac{rh}{n}, \text{ and } NM = \frac{h}{n},$$

the line AD being divided into n equal portions.

Then, drawing PP' through N parallel to the base BC,

$$PP' = 2 \frac{rh}{n} \tan \frac{A}{2}, \text{ and } z = c + \frac{rh}{n},$$

$$\Sigma (z^2 a) = \Sigma \left(c + \frac{rh}{n} \right)^2 2r \frac{h^2}{n^2} \tan \frac{A}{2}.$$

Taking the sum from $r = 1$ to $r = n$,

$$\Sigma (z^2 a) = 2 \frac{h^2}{n^2} \tan \frac{A}{2} \left\{ c^2 \Sigma (r) + \frac{2ch}{n} \Sigma (r^2) + \frac{h^2}{n^2} \Sigma (r^3) \right\}.$$

Now
$$\Sigma (r) = \frac{1}{2} n (n+1), \quad \Sigma (r^2) = \frac{n^2}{2} + \frac{n^3}{3} + \frac{n}{6}$$

and
$$\Sigma (r^3) = \left\{ \frac{n (n+1)}{2} \right\}^2 ;$$

$$\therefore \qquad \Sigma(z^2a)=2\,\frac{h^2}{n^2}\tan\frac{A}{2}\left\{c^2\,\frac{n^2+n}{2}+2ch\left(\frac{n}{2}+\frac{n^2}{3}+\frac{1}{6}\right)+h^2\,\frac{\overline{n+1,}^{2}}{4}\right\}$$

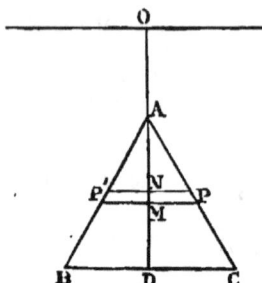

and making n infinite this becomes

$$2h^2\tan\frac{A}{2}\left(\frac{c^2}{2}+\frac{2}{3}\,ch+\frac{h^2}{4}\right).$$

Also $\Sigma(waz)=$ the whole pressure

$$=wh^2\tan\frac{A}{2}\left(c+\frac{2}{3}\,h\right);$$

$\therefore$ the depth of the centre of pressure

$$=\frac{c^2+\frac{4}{3}\,ch+\frac{h^2}{2}}{c+\frac{2}{3}\,h}=\frac{6c^2+8ch+3h^2}{6c+4h}.$$

181. In the third Example of Art. 55 the actual line of action of the fluid pressure may be found by a geometrical process.

Suppose OV, the altitude of the cone, divided into small equal parts NN', and let horizontal planes through the points of division mark out the surface of the semi-cone into a number of semicircular rings.

Let PN be the radius of one of these rings; then the pressure at every point of the ring, and therefore the resultant pressure upon the ring, passes through the point F in the axis, PF being the normal at P.

Moreover, the pressure upon the ring $\propto ON$ (surface of ring)

$$\propto ON.PN,$$

$$\propto ON.NV.$$

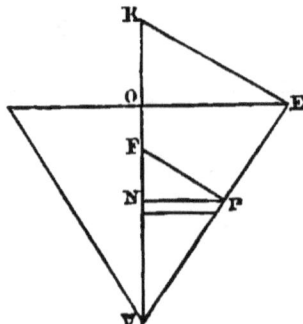

But if EK be the normal at E,

$$O.N. NV \propto EP. PV,$$
$$\propto KF. FV.$$

Upon KV as diameter describe a sphere, and let FQ be the ordinate of the sphere perpendicular to KV;

$$\therefore \qquad KF. FV = FQ^2,$$

and the pressure on the ring $\propto FQ^2$.

Hence we have to find the centre of a number of parallel forces acting at all points of KF and proportional to the areas of the sections of the sphere passing through those points.

This is clearly the same as the centre of gravity of the sphere, and it is therefore the middle point of KV.

The line of action RS therefore passes through this middle point R in the direction given by the equation

$$\tan \theta = \frac{\pi}{2} \tan a,$$

where θ is the inclination of RS to the horizon.

S is therefore the centre of pressure.

To find its position, we have

$$\tan \theta = \frac{RM}{SM} = \frac{RV - MV}{MV. \tan a};$$

$$\frac{RV}{MV} - 1 = \frac{\pi}{2} \tan^2 a;$$

$$MV = \frac{RV}{1 + \frac{\pi}{2} \tan^2 a}.$$

But $\qquad RV = \frac{1}{2} KV = \frac{1}{2} EV \sec a;$

$$SV = MV \sec a = \frac{\frac{1}{2} EV \sec^2 a}{1 + \frac{\pi}{2} \tan^2 a}.$$

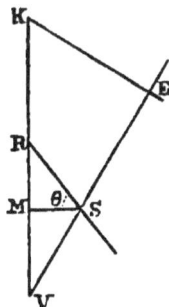

182. *One asymptote of an hyperbola lies in the surface of a fluid; it is required to find the depth of the centre of pressure of the area included between the immersed asymptote, the curve, and two given horizontal lines in the plane of the hyperbola.*

Taking OA, OB as the axes, let PN, $P'N'$ be two lines near each other and parallel to OA.

The pressure on the small area PN'

$$= wON \sin \omega. PN. NN'.$$

But $ON.PN\sin\omega$ is the area of the parallelogram $OMPN$, the constancy of which is a known property of the hyperbola.

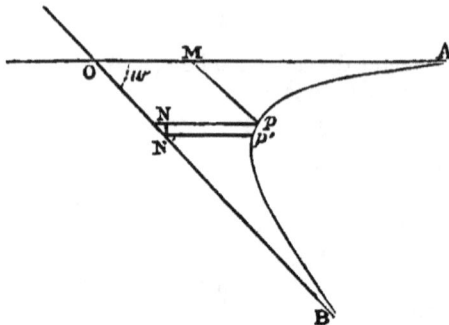

Hence the pressure on PN' varies as its vertical thickness, and therefore the depth of the centre of pressure of any finite area contained between two horizontal lines, the curve and the asymptote, is half the sum of the depths of the horizontal lines.

183. *A triangular area is immersed with one angular point in the surface; it is required to find its centre of pressure.*

Dividing the base BC into a large number of equal parts, the centre of pressure of an elementary triangle AP will be at a point R such that

$$AR = \frac{3}{4} AP,$$

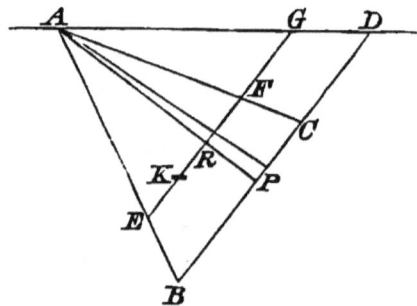

P being the middle point of the base of the elementary triangle.

If $AE = \frac{3}{4} AB$, the centre of pressure, K, of ABC will be on the line EF parallel to BC.

Further, all the elementary triangles being equal, the pressure on AP will be proportional to the depth of its centre of gravity, and therefore will vary as RG.

Hence it follows that K is the same as the centre of gravity of the frustum EF of a triangle, vertex G, and

$$\therefore \qquad GK(GE^2 - GF^2) = \frac{2}{3}(GE^3 - GF^3),$$

or $GK = \frac{2}{3} \cdot \dfrac{GE^2 + GE.GF + GF^2}{GE + GF} = \frac{1}{2} \cdot \dfrac{BD^2 + BD.CD + CD^2}{BD + CD} \quad \therefore GE = \frac{3}{4}BD.$

If β, γ be the depths of B and C,

the depth of $K = \frac{1}{2} \dfrac{\beta^2 + \beta\gamma + \gamma^2}{\beta + \gamma}$.

We can now by the aid of Art. 50 find the depth, z, of the centre of pressure of a triangle ABC in terms of the depths a, β, γ of its angular points.

Draw a horizontal plane through A and remove the liquid above; then, if z' be the depth of the centre of pressure below A,

$$z' = \frac{1}{2} \frac{(\beta-a)^2 + (\beta-a)(\gamma-a) + (\gamma-a)^2}{\beta+\gamma-2a}.$$

Replacing the liquid, and taking S for the area, we have a new pressure wSa at the centre of gravity, and therefore

$$wS \frac{a+\beta+\gamma}{3} z = wS \frac{\beta+\gamma-2a}{3}(z'+a) + wSa \frac{a+\beta+\gamma}{3},$$

or $\qquad 2z(a+\beta+\gamma) = a^2 + \beta^2 + \gamma^2 + \beta\gamma + \gamma a + a\beta.$

If h, k, l be the depths of the middle points of the sides of the triangle,

$$z(h+k+l) = h^2 + k^2 + l^2.$$

A similar method may be employed to find the centre of pressure of a sector of a circle with its centre in the surface.

Taking the case of a sector with one bounding radius (c) in the surface, divide the sector into a large number of small triangles; the centres of pressure of these triangles will be on the arc of a circle of radius $\frac{3}{4}c$, and it can be shewn, by the summation of a trigonometrical series, that the depth of the centre of pressure is

$$\frac{3c}{16} \frac{2a - \sin 2a}{1 - \cos a},$$

$2a$ being the angle of the sector.

184. *A cylindrical vessel, open at the top, is inverted and pushed down vertically in water; the substance of the vessel being of greater density than water, it is required to prove that, at a certain depth, it will be in a position of equilibrium which for vertical displacements is unstable.*

As the vessel is forced downwards the pressure of the water compresses the air within, and there must be some depth at which the air will be so compressed that the weight of the water displaced by the vessel and the air is exactly equal to the weight of the vessel and air together. At this point there will be equilibrium; but, if the vessel be slightly lifted, the air within will expand, and the weight of water displaced will be too great for equilibrium; hence the vessel will ascend. If on the other hand it be slightly depressed, a further compression of the air will take place, and the vessel will then descend.

185. *A square lamina floats with its plane vertical, and one angular point below the surface; it is required to find its positions of equilibrium.*

Let PQ be the surface of the liquid, G the centre of gravity of the square, and H of the liquid displaced, E being the middle point of PQ.

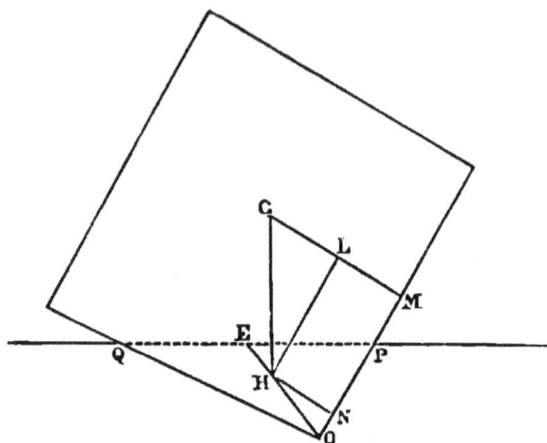

Then, if $OP=x$, and $OQ=y$, and if ρ, σ be the densities of the liquid and the lamina, and $2a$ the side of the square,

$$\frac{1}{2}\rho xy = 4\sigma a^2, \quad \text{or} \quad xy = 8\frac{\sigma}{\rho}a^2 = c^2 \text{ suppose.}$$

We have now to express the condition that GH is vertical.
Draw HN perpendicular to OP;

Then $\qquad\qquad ON = \frac{1}{3}x, \text{ and } HN = \frac{1}{3}y.$

Hence, if GM, HL be perpendicular and parallel to OP, the tangent of the angle which HG makes with OP

$$= \frac{GL}{HL} = \frac{GM-HN}{OM-ON} = \frac{a-\frac{1}{3}y}{a-\frac{1}{3}x};$$

but this angle is the complement of OPQ, of which the cotangent is $\frac{x}{y}$;

$\therefore \qquad\qquad \dfrac{3a-y}{3a-x} = \dfrac{x}{y};$

or $\qquad\qquad x^2 - y^2 = 3a\,(x-y).$

This equation gives $\qquad\qquad x=y,$

and $\qquad\qquad x+y=3a.$

The first result gives the symmetrical position of equilibrium, for which $x=y=c$.

From the second,
$$x+\frac{c^2}{x}=3a,$$

$$\therefore \qquad x=\frac{3a}{2} \pm \sqrt{\frac{9a^2}{4}-c^2}.$$

Hence, if $\frac{9a^2}{4}>c^2$, i.e. if $\frac{\rho}{\sigma}>\frac{32}{9}$, there are two other positions of equilibrium.

If $\frac{\rho}{\sigma}=\frac{32}{9}$, it will be seen that these three positions coincide.

186. *To find the vertical angle of a solid right cone on a circular base which can float with its highest generating line horizontal.*

Let the figure be a section of the cone through its axis AO, and let AC be the horizontal generating line.

G being the centroid of the cone, produce BG to D.

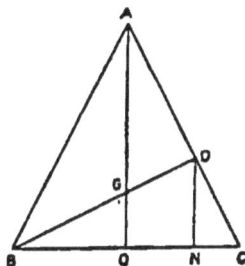

Then $\qquad \frac{DN}{OG}=\frac{BN}{BO}$, and $\frac{DN}{OA}=\frac{CN}{CO}$,

$\therefore \qquad OG.BN=OA.CN,$

and $\therefore \qquad CN=\frac{1}{4}BN.$

Hence $\qquad ON=\frac{3}{5}OC,$

and $\therefore \qquad AD=\frac{3}{5}AC.$

This shews (see page 3) that the centroids of all the parabolic sections parallel to AC, and perpendicular to the plane AGC, lie on the line BGD, and therefore that the centroid of the displaced liquid lies on that line.

Hence, whatever be the density of the liquid, as long as it is greater than the density of the cone, all that is necessary for equilibrium in the assigned position is that BGD should be a right angle.

If then θ is the semivertical angle,

$$\cos 2\theta=\frac{AD}{AB}=\frac{3}{5}, \quad \text{or} \quad \operatorname{cosec} \theta=\sqrt{5}.$$

187. *A vessel in the form of a paraboloid is immersed with its open end downwards, in a trough of mercury. Supposing the length of the axis of the vessel to be to the height of the barometer as 45 is to 64, it is required to find the depth of the surface of the mercury within the vessel when the whole vessel is just immersed.*

Let AM be the height of the vessel, and h the height of the barometer; then

$$AM=\frac{45}{64}h.$$

12—2

If PN be the surface of the mercury within the vessel, and Π' the pressure of the air within,

$$\frac{\Pi'}{\Pi} = \frac{\text{volume } AQM}{\text{volume } APN} = \frac{AM^2}{AN^2};$$

but $\qquad\qquad \Pi' = \Pi + w \,.\, AN, \text{ and } \Pi = wh;$

$\therefore$, if $\qquad\qquad\qquad AN = x,$

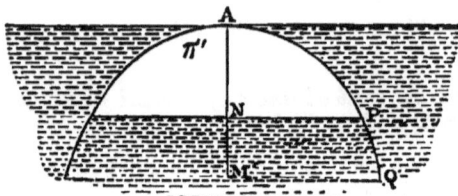

$$\frac{h+x}{h} = \left(\frac{45}{64}\right)^2 \frac{h^2}{x^2},$$

or $\qquad\qquad\qquad x^3 + hx^2 = \left(\frac{45}{64}\right)^2 h^3.$

Writing $\dfrac{z}{16}$ for $\dfrac{x}{h}$, this becomes

$$z^3 + 16z^2 = 45^2,$$

from which we find easily by trial $z = 9$, and that this is the only real root,

and $\therefore$ $\qquad\qquad\qquad AN = \frac{9}{16} h.$

188. A cylindrical vessel contains a given quantity of fluid. In this fluid is placed another cylindrical vessel of half the diameter of the first and containing half the quantity of fluid which is of half the specific gravity of that in the first vessel. In this second vessel is placed a third related to the second as the second is to the first; and so on indefinitely. Find the distance between the surfaces of the first and n^{th} fluids, neglecting the weights of the vessels.

Let w, $\dfrac{1}{2}w$, $\dfrac{1}{2^2}w$, &c. be the intrinsic weights,

r, $\dfrac{1}{2}r$, $\dfrac{1}{2^2}r$, ... the radii, and

h, h_1, h_2, ... the heights of fluid in the respective cylinders.

Then $\qquad r^2 h = 2\left(\dfrac{r}{2}\right)^2 h_1 = 2^2\left(\dfrac{r}{2^2}\right)^2 h_2 \cdots\cdots = 2^{n-1}\left(\dfrac{r}{2^{n-1}}\right)^2 h_{n-1}.$

$\therefore$ $\qquad\qquad\qquad h_{n-1} = 2^{n-1} h.$

If $\pi r^2 h = V$, the whole weight of fluid in all the cylinders beginning with the second

$$= w \left(\frac{1}{2} \frac{V}{2} + \frac{1}{2^2} \frac{V}{2^2} + \dots \text{ to infinity} \right)$$

$$= \frac{1}{3} w V.$$

This whole weight is floating in the fluid of the first cylinder, and therefore if z be the depth immersed of the second cylinder,

$$w \frac{\pi r^2 z}{4} = \frac{1}{3} w V = \frac{1}{3} w \pi r^2 h,$$

whence
$$z = \frac{4}{3} h.$$

But the effect of this immersion is to raise the surface in the first cylinder to a certain height x such that

$$\pi r^2 x - \pi \frac{r^2 z}{4} = \pi r^2 h,$$

or
$$x = \frac{4}{3} h.$$

The base of the second cylinder therefore just descends to the base of the first, and the same is the case with all the successive cylinders.

Hence the successive heights of the surfaces above the base are

$$\frac{4}{3} h, \quad \frac{4}{3} 2h, \quad \frac{4}{3} 2^2 h, \text{ &c.}$$

and the required distance is

$$\frac{4}{3} h (2^{n-1} - 1).$$

189. A straight tube $ABCD$ of small bore is bent at B and C so as to make ABC and BCD right angles, AB being equal to CD. The tube thus formed is moveable in a vertical plane about its centre of gravity, and being placed with BC horizontal and downwards, water is poured in (at A or D) so that c is the length of BA or CD occupied by the fluid. It is required to determine the condition of stability.

Let $BC = 2a$, and take b as the distance of G, the centre of gravity of the tube, from BC, and P, Q as the surfaces of the water.

Turn the tube through a small angle θ so that P', Q' are the new surfaces, and therefore

$$PP' = QQ' = a \tan \theta.$$

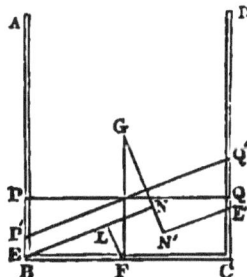

If the moment of the weight of the water about G be in the direction opposite to the displacement, the equilibrium will be stable.

Taking κ as the area of a section of the tube, this moment

$$= w\kappa \{2ab \sin \theta + (c - a \tan \theta) EN - (c + a \tan \theta) E'N'\},$$

E, E' being the middle points of $P'B$, $Q'C$; EN, EN' perpendiculars on the new vertical through G, and FL perpendicular to EN.

But $\qquad EN = LN + BF \cos \theta - EB \sin \theta$

$$= b \sin \theta + a \cos \theta - \frac{1}{2}(c - a \tan \theta) \sin \theta,$$

and $\qquad E'N' = a \cos \theta + \frac{1}{2}(c + a \tan \theta) \sin \theta - b \sin \theta.$

Hence, supposing θ very small, $\sin \theta = \theta$, $\cos \theta = 1$, and the moment

$$= w\kappa \left\{ 2ab\theta + (c - a\theta)\left(b\theta + a - \frac{1}{2}c\theta\right) - (c + a\theta)\left(a - b\theta + \frac{1}{2}c\theta\right)\right\}$$

$$= w\kappa (2ab\theta + 2bc\theta - c^2\theta - 2a^2\theta),$$

and this is positive if

$$2ab + 2bc > 2a^2 + c^2,$$

or $\qquad c^2 - 2bc + b^2 < b^2 + 2ab - 2a^2.$

If $c > b$, this leads to

$$c < b + \sqrt{b^2 + 2ab - 2a^2},$$

$c < b$, to $\qquad c > b - \sqrt{b^2 + 2ab - 2a^2}.$

If we suppose the ends A, D, joined by a continuance of the tube and the figure $ABCD$ to be a square, $b = a$, and the condition is simply

$$c < 2a,$$

so that in this case the equilibrium is always stable.

190. *Particular cases of curves of buoyancy.*

If the floating body be a plane lamina bounded, so far as regards the immersed portions, by an elliptic arc, the curves of buoyancy are similar and similarly situated concentric ellipses. This can be seen at once by projecting, orthogonally, the elliptic arc into a circle.

If the centre of gravity of the lamina is situated on the axis of the ellipse, there will be three positions of equilibrium, or only one, according as the centre of gravity is above or below the centre of curvature, at the end of the axis, of the curve of buoyancy, because three normals can be drawn to the curve of buoyancy in the former case and only one in the latter case.

Moreover this centre of curvature being the metacentre for the case in which the axis of the ellipse is vertical, it follows that in the first case the central position of equilibrium will be unstable, and therefore the other two will be positions of stable equilibrium, in accordance with the law that positions of stable and unstable equilibrium occur alternately.

If the boundary of the lamina be a parabolic arc, the curves of buoyancy are arcs of an equal parabola.

To prove this, let QQ' be the line of floatation, PV the diameter conjugate to the chord QQ', and QD the perpendicular upon PV.

The area immersed

$$= \frac{4}{3} PV \cdot QD.$$

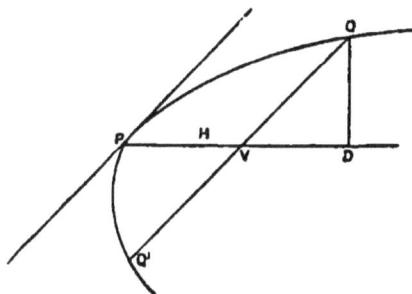

But it is a known property of the parabola* that if A is the vertex and S the focus,

$$QD^2 = 4AS \cdot PV,$$

and therefore, the area immersed being constant, it follows that QD and PV are both constant.

If H is the centroid of the displaced liquid,

$$PH = \frac{3}{5} PV,$$

and $\therefore$ PH is constant.

Hence the locus of H, which is the curve of buoyancy, is the same parabola shifted to the right.

191. *If the immersed portion of the lamina is a rectangle, we can prove that the curve of buoyancy is a parabola.*

If E is the middle point of the line of floatation PQ, any straight line through E cuts off the same area.

Take H and H' as the centroids of the displaced liquid in the two positions given by the figure, that is, when PQ and $P'Q'$ are the lines of floatation.

* See *Geometrical Conics*, Art. 46.

Then if $PQ = 2a$, and $AE = c$, and if $H'N$ is perpendicular to AE,

$$2ac \cdot HN = \frac{1}{2} a^2 \tan \theta \cdot \frac{1}{3} a \tan \theta - \frac{1}{2} a^2 \tan \theta \left(-\frac{1}{3} a \tan \theta \right)$$

$$= \frac{1}{3} a^3 \tan^2 \theta,$$

and $\quad 2ac \cdot H'N = \frac{1}{2} a^2 \tan \theta \cdot \frac{2}{3} a - \frac{1}{2} a^2 \tan \theta \left(-\frac{2}{3} a \right)$

$$= \frac{2}{3} a^3 \tan \theta.$$

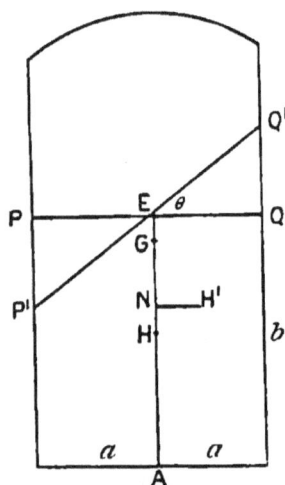

Hence $\quad\quad\quad\quad H'N^2 = \frac{2}{3} \frac{a^2}{c} HN,$

and therefore the locus of H' is a parabola.

This is a particular case of the triangular prism of Art. 69, and, as in that case, the curves of floatation and buoyancy are similar curves.

The curve of floatation is in fact a parabola with its vertex at E, and axis upwards, flattened into a straight line.

It may be remarked that, as the centre of similarity of the two curves is moved off to infinity, the visible realization of this case as the limiting case would require the application of a very powerful geometrical microscope.

Since the latus rectum of the curve of buoyancy is $2a^2/3c$ it follows that the radius of curvature at H is $a^2/3c$, and therefore there are three positions of equilibrium, or only one, according as HG is greater or less than $a^2/3c$. Further the centre of curvature being the metacentre, the central position of equilibrium in the first case will be unstable, and the other two will be positions of stable equilibrium.

It may be useful to the student to work out this case without the aid of the curve of buoyancy.

Thus, if K and L are the centroids of the triangles EQQ', EPP', and if HN, KM, LM' are perpendicular to the horizontal line through G, and if $GA = b$, and $\theta =$ the small angle QEQ', the moment about G, tending to turn the rectangle back to its original position,

$$= w\left(\frac{1}{2}a^2\theta \,.\, GM + \frac{1}{2}a^2\theta \,.\, GM' - 2ac \,.\, GN\right).$$

But $$GM = \frac{2}{3}a - EG \,.\, \theta,$$

$$GM' = \frac{2}{3}a + EG \,.\, \theta,$$

and $$GN = HG \,.\, \theta\,;$$

∴ the restorative moment is equal to

$$w\left(\frac{2}{3}a^3\theta - 2ac\theta \,.\, HG\right),\ \text{which is positive if}\ HG < \frac{a^2}{3c}.$$

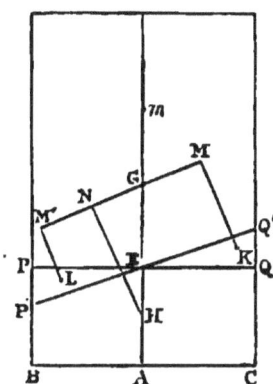

NOTE ON ART. 87.

The value of a is very nearly the same for all gases, and moreover remains nearly the same for different pressures. M. Regnault has investigated the values of a for different substances; for instance, between $0°$ and $100°$ he finds the value of a for carbonic acid gas to be ·003689. It has also been observed that the coefficients for two gases separate more from each other when the pressure is very much increased.

Regnault's results : values of a for

Air	·003665.
Hydrogen	·003667.
Nitrogen	·003668.
Sulphuric Acid	·003669.
Hydrochloric Acid	·003681.
Cyanogen	·003682.
Carbonic Acid	·003689.

NOTE ON ART. 158.

The following are approximate values of the specific heats of a few substances.

Water	1·
Thermometer-glass	·198
Iron	·114
Zinc	·1
Mercury	·03
Silver	·06
Brass	·09.

SPECIFIC GRAVITIES.

The specific gravity of Water at 60° F. is taken to be the unit.

Diamond	3·52	Nickel	8·38
Sulphur	2·	Iron	7·844
Iodine	4·94	Flint-glass	3·33
Arsenic	5·959	Plate-glass	2·5
Gold	19·4	Marble	2·716
Platina	21·53	Rock-salt	1·92
Silver	10·5	Ivory	1·917
Mercury	13·568	Ice (at 0°)	0·926
Copper	8·85	Sea-water	1·027
Tin	7·285	Olive-oil	0·915
Lead	11·445	Alcohol	0·794
Zinc	6·862	Æther	0·724

Ratios of the densities of gases and vapours of different substances to that of atmospheric air at the same temperature and under the same pressure.

Oxygen	1·103	Water	0·62
Hydrogen	0·069	Alcohol	1·613
Nitrogen	0·976	Carbonic Acid	1·524
Chlorine	2·44	Ammonia	0·591
Bromine	5·395	Sulphurous Acid	2·212
Iodine	8·701	Sulphuric Acid	2·763
Arsenic	10·365	Æther	2·586
Mercury	6·978		

1. A heavy rope, the density of which is double the density of water, is held by one end, which is above the surface, the other end being under water; find the tension at the middle point of the immersed portion of rope.

2. A triangle ABC is immersed in a fluid, its plane being vertical, and the side AB in the surface. If O be the centre of the circumscribing circle, prove that pressure on triangle OCA : pressure on triangle OCB :: $\sin 2B$: $\sin 2A$.

3. Water is gently poured into a vessel of any form; prove that when so much water has been poured in that the centre of gravity of the vessel and water is in the lowest possible position, it will be in the surface of the water.

4. If the cone be placed on its side on a horizontal table, compare the whole pressures on the curved surface and the base.

5. A triangle ABC has its plane vertical and the side AB in the surface of a liquid in which the triangle is immersed; divide it by straight lines drawn from A into n triangles on each of which the pressure shall be the same.

6. A solid displaces $\frac{1}{2}, \frac{1}{3}$ and $\frac{1}{4}$ of its volume respectively when it floats in 3 different fluids; find the volume it displaces when it floats in a mixture formed, 1st, of equal volumes of the fluids, 2nd, of equal weights of the fluids.

7. A float is made by attaching to a hemisphere (radius r) a cone of the same base, and axis of length $2r$. If this will float in a fluid A with the cone just immersed, and in a fluid B with the hemisphere just immersed, compare the densities of A and B.

8. If mercury is gradually poured into a vessel of any form containing water, prove that the centre of gravity of the mercury and water will be in its lowest position when its height above the common surface bears to the depth of water the ratio of the density of water to that of mercury.

9. A cylinder of density 2ρ floats with its axis vertical between two liquids of densities ρ and 3ρ, its height being equal to the depth of the upper liquid ; prove that the pressures on its ends are in the ratio of 1 to 5.

10. A triangular area is wholly immersed in a liquid with one side in the surface. Prove that the horizontal straight line in the plane of the area through its centre of pressure divides it into two portions, the pressures upon which are equal.

11. Shew that the centre of pressure of a parallelogram immersed with one angular point in the surface and one diagonal horizontal lies in the other diagonal and is at a depth equal to $\frac{7}{12}$ of the depth of its lowest point.

12. A parabolic lamina floats in a liquid with its axis vertical and vertex downwards ; having given the densities, σ, ρ, and the height (h) of the parabola, find the depth to which its vertex is immersed.

13. A heavy sphere, weight W, is placed in a vertical cylinder, filled with atmospheric air, which it exactly fits. Find the density of the air in the cylinder when the sphere is in a position of permanent rest, r being the radius and h the height of the cylinder.

14. A cone, of given weight and volume, floats in a given fluid with its vertex downwards ; shew that the surface of the cone in contact with the fluid is least, when the vertical angle of the cone is

$$2 \tan^{-1} \frac{1}{\sqrt{2}}.$$

15. A hollow sphere is filled with fluid and a plane drawn through the centre divides the surface into two parts, the total normal pressures upon which are as $m : 1$; find the position of the plane and the greatest and least values of m.

16. A uniform tube is bent into the form of a parabola, and placed with its vertex downwards and axis vertical : supposing any quantities of two fluids of densities ρ, ρ' to be poured into it, and r, r' to be the distances of the two free surfaces respectively from the focus, then the distance of the common surface from the focus will be

$$\frac{r\rho - r'\rho'}{\rho - \rho'}.$$

17. If there be n fluids arranged in strata of equal thickness, and the density of the uppermost be ρ, of the next 2ρ, and so on, that of the last being $n\rho$; find the pressure at the lowest point of the nth stratum, and thence prove that the pressure at any point within a fluid whose density varies as the depth is proportional to the square of the depth.

18. A fine tube, bent into the form of an equilateral triangle with its vertex upwards and base horizontal, contains equal quantities of

two liquids, each liquid filling a length of the tube equal to a side of the triangle. Prove that the height of the surface of the lighter fluid above that of the heavier : the altitude of the triangle :: $\rho' - \rho : \rho' + \rho$, ρ and ρ' being the densities of the liquids.

19. A cylinder is filled with equal volumes of n different fluids which do not mix; the density of the uppermost is ρ, of the next 2ρ, and so on, that of the lowest being $n\rho$: shew that the whole pressures on the different portions of the curved surface of the cylinder are in the ratios

$$1^2 : 2^2 : 3^2 : \dots : n^2.$$

20. Equal volumes of n fluids are disposed in layers in a vertical cylinder, the densities of the layers, commencing with the highest, being as $1 : 2 : \dots\dots : n$; find the whole pressure on the cylinder, and deduce the corresponding expression for the case of a fluid in which the *increase* of density varies as the depth.

Also, if the n fluids be all mixed together, shew that the pressure on the curved surface of the cylinder will be increased in the ratio

$$3n : 2n + 1.$$

21. A hollow cone floats with its vertex downwards in a cylindrical vessel containing water. In the position of equilibrium the area of the circle in which the cone is intersected by the surface of the fluid bears to the base of the cylinder the ratio of $6 : 19$. Prove that, if a volume of water equal to $\dfrac{19}{8}$ ths of the volume originally displaced by the cone be poured into the cone, and an equal volume into the cylinder, the position in space of the cone will remain unaltered.

22. A body is wholly immersed in a liquid and is capable of motion about a horizontal axis. It is found that the total pressure of the fluid on the surface is increased by A when the body is turned through one right angle, and further increased by B when it is turned through another right angle. Prove that the difference between the greatest and least pressures on the surface is $\sqrt{2(A^2 + B^2)}$.

23. A frustum of a right cone, formed by a plane parallel to the base and bisecting the axis, is closed and filled with fluid by means of a thin vertical pipe, which is also filled. If the top of this pipe be on a level with the vertex of the cone, find the whole pressure on the curved surface, and if this bear to the pressure on the base the ratio of 7 to 6, find the vertical angle of the cone.

24. If in the last example the base be removed, and the vessel then placed on a horizontal plane, and filled to the top of the pipe, find the least weight of the vessel which will prevent its being lifted.

25. An open cylindrical vessel, axis vertical, contains water, and a cone the radius of which is equal to that of the cylinder is placed in the water vertex downwards. Prove that, in the position of equili-

brium, if the density of the cone be one-eighth of the density of water, the surface of the water will be raised above its original level through a height equal to one-twenty-fourth the height of the cone.

26. A solid cone of wood (density σ) rests with its base on the plane base of a large vessel, and water (density ρ) is then poured in to a given height; B a piece of the same wood is then attached by a string to the vertex of the cone so as to be wholly immersed; find what the size of the piece must be in order that it may just raise the cone.

27. An elliptic lamina floats with its plane vertical in a liquid of twice the density of the lamina, 1st, with its major axis vertical, 2ndly, with its major axis horizontal; determine in each case whether the equilibrium is stable or unstable, the lamina being displaced in its own plane.

28. A regular tetrahedron has one of its faces removed and is filled with fluid; the other faces, which are capable of moving round the lowest point, are kept together by means of strings which join the middle points of the horizontal edges of the vessel; shew that the tension of the strings is to the weight of the fluid as $\sqrt{3}$ to $4\sqrt{2}$.

29. A number of weights of different densities are attached to points of a thin weightless rod. Find the density of the fluid in which it is possible for them to rest, when all are totally immersed.

If there be three weights W_1, W_2, W_3, of densities ρ_1, ρ_2, ρ_3, respectively, and x, y be the distances of W_1, W_3 from W_2, the middle weight, shew that, in order that the system may rest in equilibrium in any position when totally immersed in the corresponding fluid, the following condition must hold true,

$$\frac{x}{W_3}\left(\frac{1}{\rho_2}-\frac{1}{\rho_1}\right)+\frac{y}{W_1}\left(\frac{1}{\rho_3}-\frac{1}{\rho_2}\right)=\frac{x+y}{W_2}\left(\frac{1}{\rho_1}-\frac{1}{\rho}\right).$$

30. Two heavy liquids rest in equilibrium, one on the top of the other; one extremity of a heavy rod of length (a) is fixed at a given depth (c) in the lower liquid, and the other end reaches into the upper liquid. Find the positions of equilibrium, and determine whether they are stable or unstable.

31. A glass cylindrical vessel is inverted and plunged into water; by inclining the vessel half the air is allowed to escape, and the cylinder is then held vertically with the open end immersed and raised until one-fourth only of its length is below the surface; find the height of the water within.

32. A parallelogram is immersed in a fluid with a diagonal vertical, one extremity of which is in the surface of the fluid. Through this point lines are drawn dividing the parallelogram into three equal parts. Compare the pressures on these three parts; and, if P_2 be the pressure on the middle part, and $P_1 P_3$ those on the other two, prove that

$$16P_2=11\,(P_1+P_3).$$

33. If a solid right cone whose angle is $2a$ be immersed in a liquid with its vertex in the surface and axis vertical, prove that if P be the whole pressure on the curved surface and base, and P' the resultant pressure,

$$\frac{P}{P'} = \frac{2+3\sin a}{\sin a}.$$

Also, determine this ratio when the axis is inclined at an angle θ to the vertical, θ being less than the complement of a.

34. Three faces of a regular tetrahedron, which rests with the remaining face on a horizontal table, are heavy plates capable of moving about their horizontal edges. If they fit accurately and the tetrahedron be filled with fluid through a small hole at the vertex, shew that it will hold together if the ratio of the weight of each plate to the weight of the contained fluid be not less than 9 to 2.

35. A thin conical surface (weight W) just sinks to the surface of a fluid when immersed with its open end downwards; but when immersed with its vertex downwards a weight equal to mW must be placed within it to make it sink to the same depth as before. Shew that if a be the length of the axis, and h the height of a column of the fluid, the weight of which equals the atmospheric pressure,

$$\frac{a}{h} = m\{1+m\}^{\frac{1}{3}}.$$

36. A piston without weight fits into a vertical cylinder, closed at its base and filled with air, and is initially at the top of the cylinder; water being poured slowly on the top of the piston, find how much can be poured in before it will run over. Explain the case in which the height of the cylinder is less than the height of the water barometer.

37. Within a cylinder of height a, open at the top, is placed another cylinder of the same height, and half the content, closed at the top, and a quantity of mercury sufficient to fill the interior cylinder is poured into the exterior. If x and y be the distances of the surfaces in the two cylinders from the top, prove that

$$\frac{y}{x}(y-x) = h,$$

and find x and y; h being the height of the mercury barometer.

38. A plane rectangular lamina is bent into the form of a cylindrical surface of which the transverse section is a rectangular hyperbola. If it be now immersed in water so that first the transverse, secondly the conjugate, axes of the hyperbolic sections be in the surface, prove that the horizontal pressure on any the same immersed surface will be in the two cases the same.

39. A double funnel formed by joining two equal hollow cones at their vertices stands upon a horizontal plane with the common axis

vertical, and fluid is poured in until its surface bisects the axis of the upper cone. If the fluid be now on the point of escaping between the lower cone and the plane, prove that the weight of either cone is to that of the fluid it can hold as 27 : 16.

40. A square lamina $ABCD$, which is immersed in water, has the side AB in the surface; draw a line BE to a point E in CD such that the pressures on the two portions may be equal. Prove that, if this be the case, the distance between the centres of pressure : the side of the square :: $\sqrt{505}$: 48.

41. A cubical vessel, having one of its vertical sides moveable about a hinge in the base, is filled with water, the moveable side inclining inwards; prove that the tangent of its inclination to the horizon is to unity as the weight of the side is to the weight of the water contained by the vessel when the side is vertical.

42. A semicircular area is immersed in a liquid with its bounding diameter in the surface; find the pressure on any portion of the area contained between two radii, and find the area contained between the surface and a radius such that the pressure upon it may be one-fourth of the pressure upon the whole.

43. A vertical cylinder is filled with liquid; find the centre of pressure of the portion of its curved surface contained between two vertical planes through the axis.

44. Find the centre of pressure of the surface contained between two planes drawn through a radius of the top of the cylinder, and through the extremities of that diameter of the base which is perpendicular to the radius.

Also, find the centre of pressure of the same surface when the cylinder is inverted.

45. A solid, in the form of a right pyramid, the base of which is a regular polygon of n sides, is completely immersed in a liquid, with its base vertical; find the direction and magnitude of the resultant pressure on its inclined surfaces.

Solve the same question when the base is inclined to the vertical at a given angle.

46. An oblique cone on a circular base is completely immersed in water with its base vertical; find the resultant pressure on the curved surface.

47. A vessel in the form of an oblique cone on a circular base is held with its base horizontal and vertex downwards and is filled with liquid; find the resultant pressure on the surface and its point of action.

48. If a parabolic area be just immersed in water, and be turned about in a vertical plane so that the surface is always a tangent, prove that the centre of pressure of the part above a fixed horizontal plane

lies in the diameter through the point of contact and at a given distance from that point.

49. A portion of a right circular cone cut off by a plane through the axis and a plane perpendicular to the axis is immersed in fluid with the vertex in the surface, and axis vertical; shew that the resultant horizontal pressure on any part of the curved surface intercepted between two horizontal planes will pass through the centre of gravity of the intercepted portion of the cone.

50. A hollow cylinder is closed at one end and open at the other, and a fixed stop perpendicular to the axis divides the cylinder into two equal parts cutting off the communication between the parts; the weight of the whole cylinder is half the weight of the water which it would contain. Prove that if the cylinder be placed mouth downwards in water the depth of the stop in the position of rest will be only half as great as if a hole had been made in the stop.

51. If a thermometer plunged incompletely in a liquid whose temperature is required indicate a temperature t, and τ be that of the air, the column not immersed being m degrees, prove that the correction to be applied is $\dfrac{m(t-\tau)}{6480+\tau-m}$, $\dfrac{1}{6480}$ being the expansion of mercury in glass for $1°$ of temperature, assuming that the temperature of the mercury in each part is that of the medium which surrounds it.

52. A right circular cone is held in a liquid with its axis horizontal, and the highest point C of its base in the surface. Find the magnitude and direction of the resultant pressure on the curved surface, and determine the angle of the cone when the line of action of this pressure, (1) passes through C, (2) is parallel to a generating line.

53. Inside a solid sphere, formed of homogeneous substance, there is a spherical hollow, which is half filled with liquid; if the sphere rests on a horizontal plane, prove that, in the position of stable equilibrium, the spherical hollow will be in its highest position if the density of the sphere is greater than twice the density of the liquid.

54. Given the height of the water barometer and the specific gravity of mercury, find the height of the barometric column in a cylindrical diving-bell at a given depth in water.
How will this height be affected if a block of wood be floated inside the bell, first, if the wood comes from outside, secondly, if it falls from a shelf in the interior.

55. A conical vessel, having its vertex downwards, is filled with two liquids which do not mix, their common surface bisecting the axis; compare the whole pressures on the two portions of the surface.

56. A tube, in the form of an equilateral triangle, is filled with equal volumes of three liquids, the densities of which are as $1 : 2 : 3$; if the tube be held with one side horizontal, and the opposite angle

upwards, prove that the common surfaces of the liquids divide the sides in the ratio 1 : 2.

57. An isosceles triangular prism, the vertical angle of which is a right angle, floats in water with its edge horizontal, and its base above the surface, find its positions of equilibrium.

58. A cone is totally immersed in a fluid, the depth of the centre of its base being given. Prove that P, P', P'', being the resultant pressures on its convex surface, when the sines of the inclination of its axis to the horizon are s, s', s'', respectively,

$$P^2 (s' - s'') + P'^2 (s'' - s) + P''^2 (s - s') = 0.$$

59. A hollow cone without weight, filled with liquid, is suspended freely from a point in the rim of its base; prove that the total pressures on the curved surface and the base are in the ratio

$$1 + 11 \sin^2 a : 12 \sin^3 a.$$

60. A hollow cone without weight, closed and filled with water, is suspended from a point in the rim of its base; if ϕ be the angle which the direction of the resultant pressure on the curved surface makes with the vertical, and a the semi-vertical angle of the cone, prove that

$$\cot \phi = \frac{28 \cot a + \cot^3 a}{48}.$$

61. A heavy uniform chain is suspended from its two ends under water; prove that its form will be the same as if suspended in air.

62. An open conical shell, the weight of which may be neglected, is filled with water, and is then suspended from a point in the rim, and allowed gradually to take its position of equilibrium; prove that, if the vertical angle be $\cos^{-1}\frac{2}{3}$, the surface of the water will divide the generating line through the point of suspension in the ratio of 2 : 1.

63. A tube of small bore in the form of an ellipse is half filled with equal volumes of two fluids which do not mix; find in what manner the tube must be placed in order that the free surfaces of the two fluids may be the extremities of the minor axis.

64. If any curved surface, having for its base a plane area A and enclosing a volume V, be totally immersed in a fluid, find the resultant pressure on the curved surface, when the depth of the centre of gravity, and the inclination to the horizon, of the plane of the base are given.
If P_1, P_2, P_3, be these resultant pressures when the depths of the centre of gravity of the base, in a fluid of intrinsic weight w, are x, y, z respectively, and the inclinations of the base to the horizon are the same, shew that

$$P_1^2 (y - z) + P_2^2 (z - x) + P_3^2 (x - y) = w^2 A^2 (y - z) (z - x) (x - y).$$

65. A heavy chain is suspended from two points and hangs partly immersed in a fluid; shew that the curvatures of the portions just inside and just outside the surface of fluid are as $\rho - \sigma : \rho$, ρ and σ being the densities of the chain and fluid.

66. A U tube of uniform bore with open ends of equal length has mercury (s.g. 13·5) in its lower part and equal volumes of oil and water in the separate tubes above the mercury. The ends are now closed and as much mercury as would fill one inch of the tube is drawn off by a tap at the bottom. If 30 inches is the height of the mercury barometer and the lengths of the air columns in the two tubes before the mercury is removed are 2·25 inches and 2·75 inches, prove that the difference in level of the two mercury surfaces after the experiment is a little less than three-fifths of an inch.

67. Two vessels contain air having the same pressure Π but different temperatures t, t'; the temperature of each being increased by the same quantity, find which has its pressure more increased.

If the vessels be of the same size, and the air in one be forced into the other, find the pressure of the mixture at a temperature zero.

68. The temperature of the air in an extensible spherical envelope is gradually raised from $0°$ to $t°$, and the envelope is allowed to expand till its radius is n times its original length; compare the pressures of the air in the two cases.

69. A cylindrical vessel, closed at both ends, and placed so that its axis is vertical, is half filled with mercury at a temperature $0°$ C., the remaining space being occupied by air at the same temperature. The expansion of mercury between the temperatures $0°$ and $100°$ C. being ·018 of its original volume, and that of air ·3665 of its original volume for the same pressure, shew that if the temperature be raised to $20°$ C. the pressure of the air will be increased in the ratio $1·0772 : 1$.

70. The specific gravity of mercury compared with that of water at $68°$ is $13·568$ and at $212°$ is $13·704$. If the expansion of mercury between these points be $\dfrac{1}{69}$ th of its volume at the lower temperature, find that of water between the same points.

71. A hemispherical bowl is filled with water; if the internal surface be divided by horizontal planes into n portions, on each of which the whole pressure is the same, and h_r be the depth of the rth of these planes, prove that, $h_r \sqrt{n} = a \sqrt{r}$, a being the radius.

72. If a lamina in the form of a regular hexagon be immersed in liquid with one side in the surface, the depth of its centre of pressure is to the depth of its centre of gravity as 23 to 18.

73. Find the centre of pressure upon a portion of a vertical cylinder containing liquid, the portion being such as when unwrapped to form an isosceles triangle, the base of which when forming part of the cylinder is horizontal, and the vertex at the surface of the fluid.

74. A hollow cone open at the top is filled with water; find the resultant pressure on the portion of its surface cut off, on one side, by two planes through its axis inclined at a given angle to each other; also determine the line of action of the resultant pressure, and shew that, if the vertical angle be a right angle, it will pass through the centre of the top of the cone.

75. Two equal light spheres of the same substance are attached by strings of lengths r, r' to a point in the bottom of a vessel of water—they are mutually repulsive and rest at a distance x from each other: shew that the line joining them is inclined to the horizon at

$$\sin^{-1} \frac{r^2 - r'^2}{x \sqrt{2(r^2 + r'^2) - x^2}};$$

also if $\phi(x)$ be the repulsion

$$\phi(x) = \frac{Px}{\sqrt{2(r^2 + r'^2) - x^2}},$$

P being the excess of the fluid pressure over the weight of the sphere.

76. A cylindrical tube, containing air, is closed at one extremity by a fixed plate, the other extremity being open; a piston just fitting the tube slides within it, and the centres of the plate and piston are connected by an elastic string, the modulus of elasticity of which is equal to the atmospheric pressure on the piston; prove that, if l be the natural length of the string, and a its length when the air between the piston and the fixed plate is in its natural state, l being less than a, the length of the string in the position of equilibrium will be $(la)^{\frac{1}{2}}$.

77. If the depths of the angular points of a triangle below the surface of a fluid be a, b, c, shew that the depth of the centre of pressure below the centre of gravity is

$$\frac{(b-c)^2 + (c-a)^2 + (a-b)^2}{12(a+b+c)}.$$

78. Given that the centre of pressure of a disc of radius r, with one point in the surface, is at a distance p from the centre, prove that for a disc of radius R wholly immersed with its centre at a distance h from the surface, the distance between the centre of the circle and the centre of pressure is $pR^2 \div hr$.

79. If an air-pump be fitted with a barometer gauge of small section κ, and length l, prove that at the end of the first stroke the mercury will have risen a height

$$\frac{Bh}{A+B} \left\{ 1 - \kappa \frac{Ah + (A+B)l}{(A+B)^2} \right\} \text{ nearly };$$

h being the height of the barometer.

80. A hemispherical shell is floating on the surface of a liquid, and it is found that the greatest weight which can be attached to the rim

is one-fourth of the weight of the hemisphere; prove that the weight of the liquid which would fill the hemisphere bears to the weight of the hemisphere the ratio of

$$25\sqrt{5} : 20\sqrt{5} - 28.$$

81. A cylindrical diving-bell fully immersed is in equilibrium without a chain. Shew that if the exterior atmospheric pressure increase slightly, the ratio of the distance moved through by the bell if free to that moved through by the surface of the water in the bell when held fixed is $Hh + x^2 : x^2$ approximately; where H is the height of the water barometer, h the height of the bell, and x the length of that part of it which is filled with air.

82. A pyramid on a square base floats with its vertex downwards and base horizontal in a liquid. The pyramid is bisected by a vertical plane perpendicular to two sides of the base, and the two parts are connected at the vertex by a hinge. Prove that the parts will remain in contact if the ratio of the density of the pyramid to that of the liquid exceed

$$\left(\frac{3a^2}{2h^2 + 3a^2}\right)^3,$$

where h is the height and $2a$ the side of the base.

83. If a plane regular pentagon be immersed so that one side is horizontal and the opposite vertex at double the depth of that side, prove that the depth of the centre of pressure upon the pentagon is

$$a(29 + 3\sqrt{5}) \div 48,$$

where a is the depth of the lowest vertex.

84. If a quadrilateral lamina $ABCD$ in which AB is parallel to CD be immersed in water with the side AB in the surface, the centre of pressure will be at the point of intersection of AC and BD if

$$AB^2 = 3 . CD^2.$$

85. Supposing a common hydrometer immersed in a liquid less dense than water as far as the point to which it would sink in water, prove that, if let go, it will sink through a distance $2w(1 - s)/ks$, w being the weight of the hydrometer, k the section of its stem, and s the specific gravity of the liquid, and the hydrometer being supposed never to be entirely immersed.

86. A hollow paraboloidal vessel floats in water with a heavy sphere lying in it, there being an opening at the vertex; the water occupies the whole of the space between the vessel and the sphere. If the resultant pressure on the sphere be equal to half the weight of the water which would fill it, shew that the depth of the centre of the sphere below the surface of the water is $4a^2/3c$ where $4a$ is the latus rectum of the paraboloid, and c the distance of the plane of contact from the vertex.

87. Assuming that the weight of 100 cubic inches of air is 33 grains, when the height of the barometer is 30 inches, find the weight of the air that leaves a room when the barometer falls one inch, the room being 20 feet long 15 feet wide and 14 feet high.

88. A semicircle is immersed vertically in liquid with the diameter in the surface; shew how to divide it into any number of sectors, such that the pressure on each is the same.

89. A hollow closed vessel, in the shape of a cylinder surmounted by a cone, is filled with fluid. If the axis of the cone be three times as long as the axis of the cylinder, shew that the resultant pressure on the surface of the cone will be the same in the two positions in which the vessel can be placed with its axis vertical.

90. A small balloon containing air is immersed in water and has 100 grains of lead attached to it, the envelope of the balloon being of the same density as the water. If at the temperature of the water and the pressure of the atmosphere the balloon contain 1 cub. inch of air, find the depth to which it must be immersed in the water in order to be in a position of unstable equilibrium when the height of the water barometer is 33 feet, it being given that the density of air : that of water : that of lead as 1 : 800 : 9120.

91. If the reading of a common hydrometer when placed in fluid at the same temperature as itself be x, and if, when it is placed in the same fluid at a higher temperature than itself, its reading be at first x_1, but afterward the reading rise to x_2, the ratio of the expansions of the fluid and of the hydrometer for the same change of temperature is

$$x - x_1 : x_2 - x_1.$$

92. A solid hemisphere of radius a and weight W is floating in liquid, and at a point on the base at a distance c from the centre rests a weight w: shew that the tangent of the inclination of the axis of the hemisphere to the vertical for the corresponding position of equilibrium, assuming the base of the hemisphere entirely out of the fluid, is

$$\frac{8}{3} \frac{c}{a} \frac{w}{W}.$$

93. A cylinder has one end rounded off in the form of a hemisphere, the other pointed in the form of a cone; it floats with its axis vertical in three fluids of which the specific gravities are as $3 : 2 : 1$, in the first case the base of the hemisphere, in the last the base of the cone, is in the surface, find in what ratio the plane of floatation divides the cylindrical part in the second case.

94. A flexible and elastic cylindrical tube is placed within a rigid hollow prism, in the form of an equilateral triangle, which it just fits when unstretched; if there be no air between the tube and the prism, and if air at a given pressure be forced into the tube, find the extension and the portion in contact with the sides of the prism.

95. The readings of a perfect mercurial barometer are a and β, while the corresponding readings of a faulty one, in which there is some air, are a and b : prove that the correction to be applied to any reading c of the faulty barometer is

$$\frac{(a-a)(\beta-b)(a-b)}{(a-c)(a-a)-(b-c)(\beta-b)}.$$

96. Nicholson's Hydrometer is used to determine the weight and specific gravity of a solid and W and σ are the results when the effect of the air is neglected; prove that the actual weight is $W\{1+a/\sigma(1-a)\}$ $\{1-a/\rho\}$, where a and ρ are respectively the specific densities of air and of the material of the known weights employed.

97. A hollow vertical polygonal prism, open at both ends, rests upon a horizontal plane, every two contiguous faces being moveable about their common edge. Supposing the prism to be in equilibrium when filled with liquid ; prove that

$$\frac{c_1}{\sin a_1}=\frac{c_2}{\sin a_2}=\frac{c_3}{\sin a_3}= \dots$$

$a_1, a_2,$...being the angles of a transverse section $A_1A_2A_3...A_nA_1$, and $c_1, c_2, c_3,$...denoting the lines $A_nA_2, A_1A_3, A_2A_4,$...

98. A bridge of boats supports a plane rigid roadway AB in a horizontal position. When a small moveable load is placed at G the bridge is depressed uniformly ; when the load is placed at a point C the end A is unaltered in level; when at D the end B is unaltered in level ; and when at P the point Q of the roadway is unaltered in level.

Prove that $AG \cdot GC=BG \cdot GD=PG \cdot GQ$, and that the deflection produced at a point R by a load at P is equal to the deflection produced at P by the same load at R.

99. A thin conical shell, vertical angle $2a$, is bounded by a plane inclined at an angle θ to the axis of the cone, and is closed by an elliptic lamina of the same substance as the shell. If the shell is now held under water with the axis of the cone horizontal, prove that the whole pressures on the curved surface and on the elliptic base are in the ratio of $\sin \theta$ to $\sin a$.

Prove also that if a heavy particle, the weight of which is to the weight of the shell and its base together in the ratio of $\tan a$ to $\tan \theta - \tan a$, is attached to that point of the elliptic base which is nearest the vertex, the shell will float with the axis of the cone vertical, and the elliptic base above the surface, in any liquid the density of which exceeds a certain determinable density.

100. Two equal uniform rods AB, AC are rigidly connected at A, and the system floats symmetrically with the point A downwards.

If a is the length of each rod, and c the length of each immersed, prove that the equilibrium will be stable for a small angular displacement in the vertical plane of the rods if

$$c(3-\cos \omega) > a(1+\cos \omega),$$

where ω is the angle BAC.

CHAPTER XIII.

192. THROUGHOUT the whole of the preceding Chapters we have employed, as the unit of force, the weight of a pound at the place of observation or at some standard place.

For the next two Chapters we shall find it convenient to adopt a different unit of force.

The British Absolute Unit of Force is defined to be that force which, acting for one second upon a mass of one pound, produces in that mass the velocity of one foot per second.

In other words it is the force which produces, when acting on a mass of one pound, the unit of acceleration, taking a foot and a second as the units of length and time.

The relation between force, mass, and acceleration is therefore $P = Mf$, and it follows that if W is the weight of a mass M,

$$W = Mg.$$

The weight of a pound therefore is equivalent to g units of force, so that the unit of force is roughly equal to the weight of half an ounce.

This absolute unit of force is called the Poundal.

If the number of poundals equivalent to a force be known, the number of pounds weight to which the force is equivalent will be obtained if we divide the number of poundals by the value of g, referred to a foot and a second as units, at the place at which the force is in action.

193. If ρ is the density of a substance, and M the mass

of a volume V, $M = \rho V$, and therefore if W is the weight of the volume V,

$$W = g\rho V \text{ poundals.}$$

It will be seen that if we change the locality W and g are changed in the same proportion, so that the equation is true at any place, although the number of poundals in the weight of the mass ρV is dependent upon locality.

The value of g at the equator, when a foot and a second are units of length and time, is 32·088, and at a place of latitude λ, g is given by the equation

$$g = (32\cdot088)\ \{1 + \cdot005133 \sin^2 \lambda\},$$

so that the value of g at the North or South Pole, would be 32·2527.*

For ordinary calculations, not requiring great accuracy, the value of g usually adopted is 32·2, this being approximately the value in London and in Paris.

194. If in the preceding Chapters we had employed the British absolute unit of force, the expressions for pressures, whole pressures and resultant pressures would have been slightly different in form, the difference being that $g\rho$ would appear in the place of w.

Thus, in Chapter III., if p is the pressure, in poundals, at the depth z, we should have

$$p = g\rho z,$$

and the expression for whole pressure, in poundals, would be

$$g\rho \bar{z} S.$$

Also, in Art. 57, the expressions for the resultant horizontal and resultant vertical pressure in poundals, would be

$$g\rho A z \cos \theta, \text{ and } g\rho \ \{V + Az \sin \theta\}.$$

Again, in Art. 77, the expression for atmospheric pressure at the height z would be

$$\Pi - g\rho z.$$

Finally, it must be noticed that the meaning of k in the equation, $p = k\rho_0$, will be changed.

---

* Thomson and Tait's *Natural Philosophy*, Part I., page 226.

Since the pressure of a given kind of gas, of a given density, and at a given temperature, is independent of time and place, it follows that when the pressure is measured in absolute units of force, the quantity k is an absolute constant, independent of time and place.

If then h is the height of the homogeneous atmosphere at a place at which ρ_0 is the density of the air, when the temperature is $0°$ C., $g\rho_0 h = k\rho_0$, so that k is g times the height, in feet, of the homogeneous atmosphere.

Conversely the height of the homogeneous atmosphere, at any place, is the quotient obtained when we divide the numerical value of k by that of g at the place.

If t is the temperature, the equation is

$$g\rho h = k\rho\,(1 + \alpha t),$$

so that $k\,(1 + \alpha t)$ is g times the height of the homogeneous atmosphere.

Numerical value of k for atmospheric air.

Taking ·0013 and 13·606 as the specific gravities of air and mercury at the temperature $0°$ C., and 30 inches as the height of the barometer, the equation, $g\sigma h = k\rho$, becomes

$$32{\cdot}2 \times 13{\cdot}606 \times 62{\cdot}5 \times 2{\cdot}5 = k\,({\cdot}0013) \times 62{\cdot}5,$$

the unit of mass being a pound and the unit of force a poundal.

We hence find that the value of k is about 840000.

The value of the absolute constant R in the equation, $pV = RT$, must in a similar manner be determined, for any particular gas, by experimental observations.

The C.G.S. system.

195.   In France the system of units adopted is the Centimetre-Gramme-Second system.

Taking a gramme for the unit of mass, the unit of force is the force which produces, in one second, when acting on a gramme the velocity of one centimetre per second. This is the French absolute unit of force and is called the *Dyne*.

In this case the equation,

$$W = Mg,$$

asserts that the weight of M grammes is Mg dynes.

If the number of dynes equivalent to a force be known, the number of grammes weight to which the force is equivalent will be found if we divide the number of dynes by the value of g, referred to a centimetre and a second as units of length and time, at the place at which the force is in action.

At the equator the value of g, referred to a foot and a second is 32·088, and therefore, since one foot contains 30·4797 centimetres, the value of g, referred to a centimetre and a second is, approximately, 978·0326.

Hence it follows that, in latitude λ,

$$g = (978 \cdot 0326) \cdot \{1 + \cdot 005133 \sin^2 \lambda\}.$$

For ordinary calculations the value of g which is usually employed is 981, which is very nearly its value in Paris and in London.

196.   The remarks of Art. 194 are equally applicable to the case in which the French absolute unit of force is employed.

The expression $g\rho z$, for the pressure at the depth z, gives in dynes, the pressure on a square centimetre; and the expression $g\rho \bar{z} S$ gives the whole pressure, in dynes, on the surface S, measured in square centimetres.

Also, if ρ is the density and p the pressure of a gas, the equation, $p = k\rho$, asserts that the pressure upon a square centimetre is $k\rho$ dynes, ρ being the number of grammes in a cubic centimetre of the gas.

Taking h as the height in centimetres of the homogeneous atmosphere at the place, the equation

$$g\rho h = k\rho$$

shews that k is g times the height, in centimetres, of the homogeneous atmosphere.

197.   To find the number of dynes in a poundal, observe that a pound is 453·59265 grammes, and therefore that the poundal produces, in one second, the velocity of one foot per

second, that is, of 30·4797 centimetres per second, in a mass of 453·59265 grammes.

Hence it follows that a poundal, acting for one second on a mass of one gramme, would produce the velocity, in centimetres per second, measured by the number

$$(453 \cdot 59265) \times (30 \cdot 4797).$$

This is approximately equal to 13825, which is therefore, approximately, the number of dynes in a poundal.

It may be useful to place, in a tabular form, some numerical facts. Taking g to be 32·2 when a foot and a second are units,

weight of 1 lb.	=	32·2	poundals	
„ „ 1 cwt.	=	3606·4	„	
„ „ 1 ton	=	72128	„	
„ „ 1 kilogramme	=	71	„	nearly.
„ „ 1 gramme	=	·071	„	„

Taking g to be 981 when a centimetre and a second are units,

weight of 1 gramme	=	981	dynes	
„ „ 1 kilogramme	=	981000	„	
„ „ 1 grain	=	63·568	„	nearly.
„ „ 1 pound	=	445000	„	„
„ „ 1 cwt.	=	49840000	„	„

Surface Tensions of Liquids.

The following table of surface tensions is taken from Quincke's results.

The tensions are given, in degrees, for the temperature 20° C.

	Specific gravity	Tension of surface separating liquid from			Angle of contact with glass in presence of		
		air	water	mercury	air	water	mercury
Water	1	81		418	25° 32′		26° 8′
Mercury	13·543	540	418		51° 8′	26° 8′	
Chloroform	1·488	30·6	29·5	399			
Alcohol	·79	25·5		399	25° 12′		
Olive oil	·913	36·9	20·56	335	21° 50′	17°	47° 2′
Turpentine	·887	29·7	11·55	251	37° 44′	37° 44′	47° 2′

Since the poundal is approximately equal to 13825 dynes we must divide the numbers in the second, third and fourth columns by 13825, in order to obtain the tensions in poundals.

If we take Mensbrugghe's results at the end of Chapter X., and transform the gramme weights into dynes, we shall find, on comparing the two tables that Quincke's give rather higher values for surface tensions.

Work.

198. When a heavy body, lying on the ground, is lifted to a given height, it is said that work is done by the lifting force, and the measure of the work done is the product of the numerical quantities measuring the force and the height.

If a body is displaced in any direction by a force in that direction, the work done is measured by the quantity Px, where P is the force, supposed to be exerted uniformly, and x the space through which it is exerted.

If the direction of the force is not the same as the direction of displacement, the work done is the product of the displacement by the resolved part of the force in that direction.

Thus if a heavy body on a smooth inclined plane is dragged over the space x, up the line of greatest slope, by a force acting in a direction inclined to the plane at the angle θ, the work done will be the product of x, and the component of the force parallel to the plane, and therefore be represented by the expression $Px \cos \theta$.

In the British Isles, when the unit of force employed is the weight of one pound, the unit of work is the foot-pound, that is, it is the work done by a force of one pound weight exerted through one foot.

In France, when the unit of force is the weight of a kilogramme, the unit of work is the kilogramme-metre, that is, it is the work done by a force equal to the weight of one kilogramme exerted through one metre.

When we employ absolute units of force, the British unit of work is the foot-poundal, that is, the work done by one poundal exerted over one foot.

If we take the French absolute unit of force, that is, the dyne, the French unit of work is the erg, which is the work done by one dyne exerted over the space of one centimetre.

If the element of time is introduced the rate of doing work is called power.

The unit of Power employed by British Engineers is the Horse-power. The Horse-power is the rate at which an agent works which does 33000 footpounds per minute, or, 550 footpounds per second.

In the C.G.S. system, taking a *joule* to represent 10000000 ergs, the foot-pound is 1·356 joules.

In this system the unit of power is the *Watt*, which is work at the rate of one joule per second.

Taking g to be 981, when a centimetre and a second are units, the following table connects some of the various measures of work with the erg.

one gramme-centimetre	=	981 ergs.
one kilogramme-metre	= 98100000	,,
one foot-pound	= 13560000	,,
one foot-grain	=	1937 ,,
one joule	= 10000000	,,
one horse-power	=	44748 joules per minute.
,, ,, ,,	=	745·8 Watts.

EXAMPLES.

Assume that g is 32·2 when a foot and a second are units, and is 981 when a centimetre and a second are units.

1. Find in poundals per square foot, and in dynes per square centimetre, the pressure at the depth of 100 feet in a lake, (1) neglecting, (2) taking account of atmospheric pressure, assuming that 33 feet is the height of the water barometer.

2. A cubical box, the edge of which is one foot, is filled with equal volumes of olive oil and alcohol, the specific gravities of which are ·9 and ·8; calculate in poundals the pressures on the base and on each vertical side.

3. Taking the pressure of the atmosphere as equal to the weight of 14½ lbs. per square inch, find its value in dynes per square centimetre, assuming that a gramme is ·0022 of a pound, and that a metre is 39·37 inches.

4. A solid hemisphere, of radius ten centimetres, is held with its plane base vertical and just immersed in a liquid of density 13·568 grammes per cubic centimetre; find the expressions in dynes for the resultant horizontal and resultant vertical pressures on its curved surface.

5. A cubic foot of air, the pressure of which is equal to the weight of 15 lbs. per square inch, is compressed slowly into a cubic inch ; find the new pressure in poundals per square inch, and in dynes per square centimetre.

6. The height of the barometer being 30 inches, and the specific gravity of the mercury being 13·568, calculate the pressure of the atmosphere in poundals per square inch.
Also calculate the height of the barometer in centimetres, and the pressure of the atmosphere in dynes per square centimetre.

7. A cylinder formed of flexible and inextensible material, closed at both ends, contains gas at the pressure of 21·56 lbs. weight per square inch; taking 15 lbs. weight per square inch as the atmospheric pressure, find the tension, in poundals, of the curved surface of the cylinder.

8. A soap-bubble of radius two centimetres is blown from a mixture the surface tension of which is 80 dynes per centimetre, and another of radius 2·5 centimetres is blown from a mixture the surface tension of which is 30 dynes per centimetre. Compare the excesses over atmospheric pressures of the air pressures inside the bubbles. Also having given that the pressure of the atmosphere is a megadyne per square centimetre, compare the masses of air inside the bubbles.

9. A cylindrical block of wood, of height l and sectional area κ, is floating in a lake with its axis vertical. If it is pushed down very slowly until it is just immersed, prove that the work done is

$$\tfrac{1}{2}g\kappa l^2 \frac{(\rho-\sigma)^2}{\rho}$$

foot poundals, where ρ and σ denote the densities of the water and wood.

10. If the block is floating in a cylindrical vessel of sectional area κ', prove that the work done in slowly immersing it is, in foot poundals,

$$\tfrac{1}{2}g\kappa l^2 \left\{1-\frac{\kappa}{\kappa'}\right\} \frac{(\rho-\sigma)^2}{\rho}.$$

11. A cylindrical block of iron of density σ, of height l and sectional area κ, stands with its axis vertical at the bottom of a lake, at a place where h is the depth; if it is lifted slowly just above the surface, the work done is, in foot poundals,

$$g(\sigma-\rho)\kappa lh + \tfrac{1}{2}g\rho\kappa l^2.$$

CHAPTER XIV.

ROTATING LIQUID.

199. WHEN liquid in a vessel is set rotating about a vertical axis, it is known that the surface assumes a hollow form; by the help of a dynamical law we can determine what this form is.

If a mass of liquid, contained in a vessel which rotates uniformly about a vertical axis, rotates uniformly, as if rigid, with the vessel, its surface is a paraboloid.

Every particle of the liquid moves uniformly in a horizontal circle, and therefore, whatever the forces may be which act on any particle, their resultant must be a horizontal force tending to the centre of the circle, and equal to $m\omega^2 r$, where r is the distance of the particle from the axis, m is its mass, and ω is the angular velocity of the liquid*.

We assume from considerations of symmetry that the surface is a surface of revolution.

Let AG be the vertical axis of revolution, and consider a point P on the surface.

Round P as centre describe a very small circle on the surface, and take this small circular area as the base of a very thin circular cylinder of liquid.

If m be the mass of the element of liquid, thus imagined, the forces upon m will be its weight, the atmospheric pressure on its surface, the liquid pressure on the flat end, and the liquid pressures on the curved surface.

These latter pressures, being equal to the

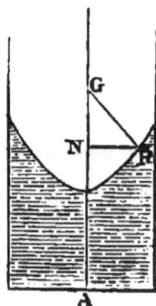

---

* See Garnett's *Dynamics*, or Loney's *Dynamics*.

atmospheric pressure, balance each other, and it therefore follows that the resultant of gravity, mg, and of the liquid pressure, which is normal to the surface, is in the direction PN, the perpendicular on the axis, and is equal to $m\omega^2 r$. Let the normal at P meet the axis in G; then, by the triangle of forces

$$NG : PN :: mg : m\omega^2 PN,$$

and $$\therefore NG = g/\omega^2.$$

Now NG is the subnormal, and it is known that in a parabola the subnormal is constant, while it can also be shewn that the parabola is the only curve which has this property.

The vertical section in the figure is therefore a parabola the latus rectum of which is $2g/\omega^2$, and the surface is a paraboloid.

It will be seen that this result is independent of the form of the containing vessel. The axis of rotation, in fact, may be within or without the fluid, but in any case it will be the axis of the paraboloidal surface.

If the vessel were a surface of revolution, having the axis of rotation for its axis, it would not be necessary theoretically that the vessel should rotate. However, by making it rotate with the liquid, we get rid of the practical difficulty which would in this case arise from the friction between the fluid and the surface of the vessel.

200. *To find the pressure at any point.*

Take any point Q in the fluid, and describe a small vertical prism having Q in its base, which is to be taken horizontal.

The prism PQ of liquid rotates uniformly under the action of the pressure around it, but its weight is entirely supported by the pressure on its base.

Hence if p be the pressure, and ρ the density,

$$p = g\rho . PQ.$$

Now

$$PQ = OM - ON = \frac{\omega^2 QN^2}{2g} - ON,$$

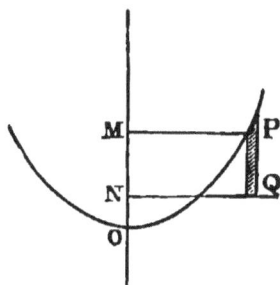

$$\text{and} \quad \therefore \quad p = \rho \left(\frac{1}{2} \omega^2 Q N^2 - g O N \right);$$

and we thus obtain the pressure in terms of the horizontal and vertical distances from the vertex of the paraboloid.

It must be observed that ON is measured upwards and that if Q be lower than O, the equation for p is

$$p = \rho \left(\frac{1}{2} \omega^2 Q N^2 + g O N \right).$$

If a foot and a second are units of length and time, this equation gives the pressure in poundals per square foot.

In order to obtain the pressure in pounds weight per square foot, we must divide by the value of g at the place.

If we take a centimetre and a second as units, the equation gives the pressure in dynes per square centimetre; and to obtain the pressure in grammes weight per square centimetre we must divide by the value of g at the place.

201. *To find the resultant vertical pressure of a rotating liquid on any surface.*

Let PQ be the surface, and draw vertical lines from its boundary to the surface; then the weight of the included portion $PABQ$ of liquid being entirely supported by PQ, it follows that the resultant vertical pressure on PQ is equal to the weight of the liquid above it.

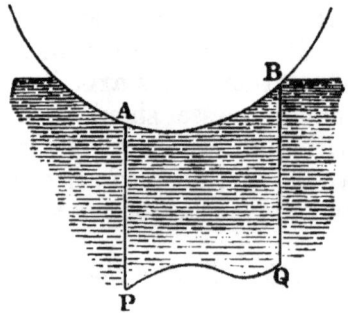

If the surface PQ be pressed upwards, as in the lower figure, then, continuing the free surface AOP of the liquid, it can be shewn, as in Art. (53), that the resultant vertical pressure upwards on PQ is equal to the weight of the fluid which would be contained between the paraboloidal surface, the surface PQ, and vertical lines through the boundary of PQ.

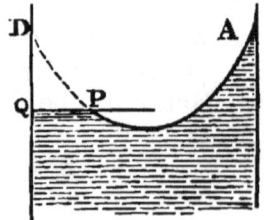

202. DEF. *A surface of equal pressure is the locus of points at which the pressures are the same.*

If lines be drawn vertically downwards from all points of the surface equal to PQ (fig. Art. 200), it is clear that the pressures at their ends will be the same as at Q; and, as these ends lie on the surface of a paraboloid equal to the surface paraboloid, it follows that all surfaces of equal pressure are in this case paraboloids.

203. *Floating bodies.* If a body float in a rotating mass of fluid, in a position of relative equilibrium, it is evident by the same reasoning, as in the case of a fluid at rest, that the weight of the body must be equal to the weight of the fluid displaced.

204. *Figure of the Earth.* A large portion of the earth's surface is covered with water, and, if it were not for the earth's rotation, its surface would be a sphere having its centre at the centre of the earth.

For simplicity, imagine a solid sphere surrounded by water, and suppose the whole to be in rotation about a diameter CB of the sphere. Consider an elementary portion P of the water, which describes a circle of radius PN uniformly. The attraction of the solid sphere in the direction PC, combined with the resultant fluid pressure in direction of the normal at P, and the attraction of the water must have as their resultant the force $m\omega^2 PN$ in direction PN. Hence the normal at P must be inclined, as in the figure, towards the axis, and the form of the surface must be oblate.

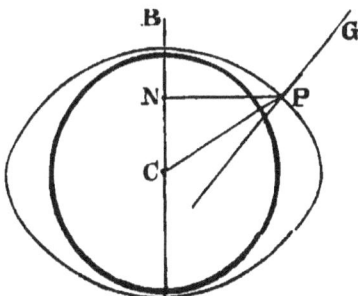

Supposing the earth a large fluid mass, it is shewn by mechanical considerations that the form would be an oblate spheroid.

It is hence seen that the normal to the surface of still water, that is, the vertical, at any point of the earth's surface is not in direction of its centre, except at the poles and the equator.

205. Regarding the Earth as a solid sphere, let P be the position of a pool of still water on its surface, and let PG be the normal to its surface.

The forces on a particle of water at P are the resultant fluid pressure in the direction of GP, and the attraction of the earth in the direction PC.

This resultant fluid pressure on the particle is the force in direction of the normal to the surface of the water which is required to maintain the particle in its position, and is therefore equal to its weight.

Since the resultant of these two forces is in the direction PN, i.e. parallel to GC, it follows that, if m is the mass of the particle,

$$m\omega^2 PN : mg :: CG : PG.$$

Let θ represent the angle PCA, ϕ the angle PGA, and ψ the angle CPG; then we obtain, if a is the Earth's radius,

$$\omega^2 a \cos\theta : g :: \sin\psi : \sin\theta.$$

Now if g is the acceleration due to gravity at the equator, $289\omega^2 a = g$, so that ψ is a very small angle, and therefore neglecting the difference between gravity at the equator and at the place,

$$\psi = \frac{\omega^2 a}{2g} \sin 2\theta = \frac{\omega^2 a}{2g} \sin 2\phi, \text{ approximately,}$$

or $\qquad \psi = \dfrac{1}{578} \sin 2\phi.$

The angle ϕ being the latitude of the place this gives the difference between the latitude and the angle PCA, which latter angle is sometimes called the geocentric latitude.

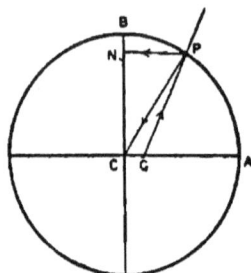

EXAMPLES.

206. (1) *A fine tube, ABCD, of which the equal branches AB, CD, are vertical, BC being horizontal, is filled with liquid, and made to rotate uniformly about the axis of AB; to find how much liquid will flow out of the end D.*

The liquid will flow out until the surface in AB is the vertex of a parabola passing through D, and having its axis vertical and latus rectum equal to $\frac{2g}{\omega^2}$.

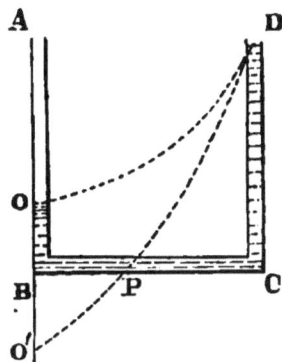

If then O be the vertex of the parabola,

$$BC^2 = \frac{2g}{\omega^2} AO.$$

This gives AO, and determines the quantity required.

If however AO be greater than AB, the surface of the liquid will be in BC, at P suppose.

In this case we have $BC^2 = \frac{2g}{\omega^2} AO'$,

$$\text{and } BP^2 = \frac{2g}{\omega^2} BO' = \frac{2g}{\omega^2}(AO' - AB),$$

which determines the position of P.

(2) *A straight tube* AB, *filled with liquid, is made to rotate uniformly about a vertical axis through* A; *to find how much flows out at* B.

Let $OAB = a$, and imagine a parabola, latus rectum $\frac{2g}{\omega^2}$, to be drawn touching the axis of the tube, and having its axis coincident with the vertical through A.

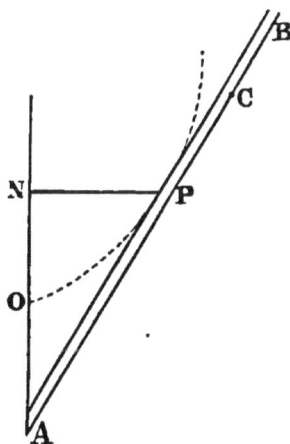

Then if P be the point of contact, all the fluid above P will flow out.

To find P,

$$PN^2 = \frac{2g}{\omega^2} . ON$$

$$= \frac{g}{\omega^2} AN, \text{ since } OA = ON,$$

and $\quad PN = AN \tan a$;

$$\therefore AN = \frac{g}{\omega^2} \cot^2 a,$$

and $\quad AP = \frac{g}{\omega^2} \frac{\cos a}{\sin^2 a}.$

No fluid will flow out unless $AP < AB$, that is, unless

$$\omega^2 > \frac{g}{AB} \frac{\cos a}{\sin^2 a}.$$

It will be seen that P is the position of relative equilibrium of a heavy particle in the rotating tube.

(3) *Let the end* B *be closed and the tube* AB, *rotating as in Ex.* (2), *be only partly filled with liquid; it is required to find the circumstances of relative equilibrium.*

Let AC be the portion of tube filled with liquid (fig. of previous article).

Draw the parabola touching in P as before : then, if C is below P, no change takes place, but if C is above P, the portion PC of liquid will flow to the upper end of the tube and remain there.

(4) *A semicircular tube* APB *is filled with liquid and rotates uniformly about the vertical diameter* AB ; *it is required to find where a hole may be made in the tube through which all the liquid will flow out.*

Draw a parabola touching the tube and having its vertex in $B.1$, axis vertical, and latus rectum equal to $\dfrac{2g}{\omega^2}$, and let P be its point of contact.

Then, if an aperture be made at P, the whole of the liquid, being above the paraboloidal surface, will flow out through P.

To find its position we have

$$PN^2 = \frac{2g}{\omega^2} \cdot ON = \frac{g}{\omega^2} NT ;$$

but $$PN^2 = CN \cdot NT ;$$

$$\therefore \ CN = \frac{g}{\omega^2}, \text{ which determines } P.$$

If a be the radius (CA) of the tube, and if $\omega^2 < \dfrac{g}{a}$, then $CN > CA$, and the aperture must be made at A.

(5) *In a mass of liquid, rotating about a vertical axis, a very small sphere, of greater density than the liquid, is immersed, and supported by a string fastened to a point in the axis; it is required to find the position of relative equilibrium.*

For one position of equilibrium it is evident that the string can be vertical, but we can shew that the sphere may rest with the string inclined at a certain angle (θ) to the vertical.

Let V be the volume of the sphere, r its distance from the axis in the position of relative equilibrium, and ρ the density of the liquid.

To find the pressure of the liquid on the sphere, imagine it removed and its place supplied by a portion V of the liquid ;

The resultant liquid pressure must support the weight $g\rho V$, and also supply the horizontal pressure necessary to maintain the circular motion, i.e. $\rho V \omega^2 r$.

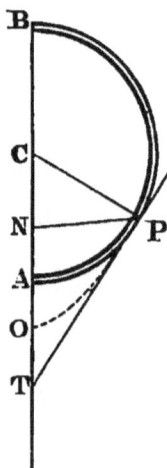

Hence if ρ' be the density of the sphere, and t the tension of the string, we must have, for equilibrium,

$$t \sin \theta + \rho V \omega^2 r = \rho' V \omega^2 r,$$

$$t \cos \theta + \rho V g = \rho' V g,$$

$$\text{and} \therefore \tan \theta = \frac{\omega^2 r}{g}.$$

The position is therefore the same as if the sphere and string were in motion as a conical pendulum.

It will also be seen that the string coincides with the direction of the normal to the surface of equal pressure which passes through the centre of the sphere.

(6) *A cylindrical vessel contains liquid, which rotates uniformly about the axis of the cylinder; to find the whole pressure on its surface.*

Let AOB be a vertical section of the surface, r the radius of the cylinder.

We have shewn, Art. (200), that the pressure varies as the depth below the surface, and in this case the level of the free surface is the same for all points on the curved surface of the cylinder.

Hence the whole pressure on the curved surface

$$= g\rho\, 2\pi r . AD . \frac{1}{2} AD = \pi g \rho r . AD^2.$$

Let h be the height of the liquid when at rest.

It is known that the volume of a paraboloid is half that of the cylinder on the same base and of the same height, and therefore the surface of the liquid at rest would bisect AN.

But
$$ON^2 = \frac{2g}{\omega^2} AN, \text{ or } r^2 = \frac{2g}{\omega^2} AN;$$

$$\therefore AD = h + \frac{1}{2} AN = h + \frac{\omega^2 r^2}{4g}.$$

Hence the whole pressure is given in terms of h.

Also the pressure on the base is equal to the weight of the fluid, i.e. $g\rho\pi r^2 h$.

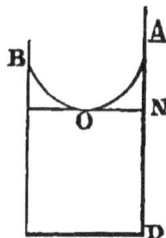

NOTES ON CHAPTER XIV.

In Art. (199), we considered the motion of a particle of water in the shape of a small thin cylinder.

The shape of the element is however quite immaterial, for an element of any shape in the surface will lie between the free surface and a consecutive surface of equal pressure, and the resultant pressure of the liquid in any direction in the tangent plane will be zero.

The Problem of rotating liquid may however be treated differently in the following manner.

Let P be a point beneath the surface of the liquid and PN its distance from the axis of revolution Az.

In the vertical plane ANP take a consecutive point Q on the surface of equal pressure passing through P, the shape of which is to be determined.

Draw the vertical line through Q, and produce NP to meet it in R.

If we imagine QR to be the axis of a very thin cylinder, it will have no vertical acceleration, and therefore the difference of the pressures at Q and R

$$= g\rho QR,$$

which is also the difference of the pressures at P and R.

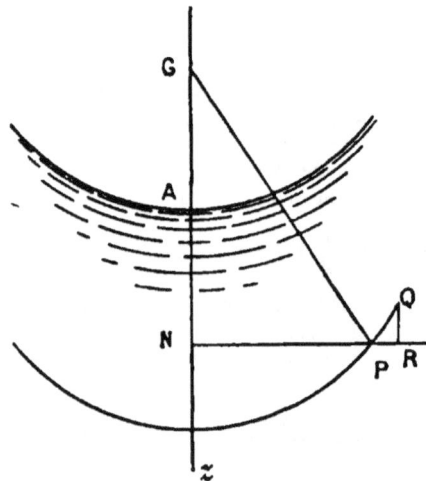

If now we imagine a very thin cylinder of which PR is the axis the mass of liquid within this cylinder has the acceleration $\omega^2 PN$ in the direction PN; and therefore, if κ is the cross section of the cylinder,

$$\rho\kappa PR \cdot \omega^2 PN = g\rho\kappa \cdot QR.$$

or $\omega^2 PR \cdot PN = g \cdot QR.$

But, if PG is the normal to the surface of equal pressure at P, the infinitesimal triangle PQR is similar to the triangle PGN, so that

$$QR : PR :: PN : NG,$$

and it follows at once that

$$NG = \frac{g}{\omega^2}.$$

Hence, the subnormal being constant, the curve is a parabola.

The surfaces of equal pressure, including the free surface, are therefore paraboloids of revolution.

The question of the form of the surface of a rotating fluid appears to have been first discussed by Daniel Bernoulli, in his *Hydrodynamica*, which was published in 1738. He there proves that the form is that of a paraboloid, and five years after Clairaut, in his *Figure de la Terre*, gives a similar proof, at the same time quoting Bernoulli.

From the remarks of Art. 17, it follows that the paraboloidal form will be exactly true for viscous liquids rotating in the same manner. The point of argument lies in the phrase, *as if rigid*, for, without this condition, it would not be possible to imagine the liquid in a state of equilibrium. It must not be inferred that the paraboloidal form is that which would be assumed by a liquid set in rotation by ordinary mechanical means. The internal friction of a liquid, communicated from the surface of a rotating vessel may produce the effect, if the revolution be maintained long enough.

Theoretically, we can imagine the effect produced by enclosing ice in a strong vessel with a paraboloidal upper surface, making it rotate, and then melting the ice by pressure, or otherwise. The melted ice would retain the rotation as if rigid, and it might perhaps be possible to procure an approximation to the paraboloidal surface.

If a cup of tea be rapidly stirred, a convex surface is produced, having a hollow in the middle, but, in motion of this kind, the angular velocity decreases at increasing distances from the centre, and there is a constant change of the relative positions of the molecules of liquid. This is the case of Rankine's free circular vortex, and its discussion belongs to the domain of Hydrodynamics. (See *Hydromechanics*.)

EXAMPLES.

1. Liquid contained in a closed vessel rotates uniformly about a vertical axis; prove that the difference of the pressures at any two points of the same horizontal line varies as the difference of the squares of the distances of the two points from the axis of rotation.

2. A hollow paraboloid of revolution with its axis vertical and vertex downwards is half filled with liquid. With what angular velocity must it be made to rotate about its axis in order that the liquid may just rise to the rim of the vessel?

3. If a solid cylinder float in a liquid which rotates about a vertical axis having its axis coincident with the axis of revolution, determine the portion of its surface which is submerged, the dimensions of the cylinder and the densities of the liquid and cylinder being given.

4. An open vessel, containing two liquids which do not mix, revolves uniformly round a vertical axis; find the form of the common surface.

5. A conical vessel open at the top and filled with liquid rotates about its axis; find how much runs over, 1st, when ω is less, and, 2nd, when ω is greater than $\sqrt{\dfrac{g}{h}}\cot a$, h being the height of the cone, and a its semivertical angle.

6. A hemispherical bowl is filled with liquid, which is made to rotate uniformly about the vertical radius of the bowl; find how much runs over.

7. An elliptic tube, half full of liquid, revolves about a fixed vertical axis in its own plane, with angular velocity ω; prove that the angle which the straight line joining the free surfaces of the liquid makes with the vertical is $\tan^{-1}\dfrac{g}{p\omega^2}$, where p is the perpendicular from the centre on the axis.

8. A closed cylindrical vessel, height h and radius a, is just filled with liquid, and rotates uniformly about its vertical axis; find the pressures on its upper and lower ends, and the whole pressure on its curved surface.

9. A hemispherical bowl, just filled with liquid, is inverted on a smooth horizontal table, and rotates uniformly about its vertical radius; find what its weight may be, in order that none of the liquid may escape.

10. A cylindrical vessel, containing water, rotates uniformly about its axis, which is vertical, the water rotating with it at the same rate; find the position of relative equilibrium of a small piece of cork which is kept under water by a string fastened to a point in the side of the vessel.

11. A vertical cylinder, of height h and radius a, is half full of water, which rotates uniformly about the axis; prove that the greatest angular velocity which can be imparted to the water without causing an overflow is $\sqrt{2gh}\div a$.

12. A conical vessel, of height h and vertical angle 60°, has its axis vertical and is half filled with water; prove that the greatest angular velocity which the water can have without overflowing is

$$\sqrt{\frac{2g}{h}}.$$

13. A fine tube whose axis is of the form of three sides of a square, each equal to a, is filled with fluid and made to rotate about a vertical axis bisecting at right angles the middle side, prove that no fluid will escape unless the angular velocity exceeds $2\sqrt{2g/a}$. If the angular velocity have this value shew that the whole pressure on the tube is two-thirds of what it would be if the fluid did not rotate at all.

14. A sphere (radius c) is just filled with water, and rotates about a vertical diameter with angular velocity ω, such that $3c\omega^2=2g$; prove that the pressure at any point of the surface of equal pressure which cuts the sphere at right angles is $3g\rho c/4$, ρ being the density of water.

15. A cylindrical vessel containing fluid is closed at the top by a heavy lid capable of moving about a hinge at a point of its circumference. If a be the radius of the cylinder, h its height, and if it be made to rotate with angular velocity $\dfrac{2\sqrt{gh}}{a}$ about its axis, which is vertical, shew that the lid will just be raised by the fluid if the weight of the fluid be to that of the lid as $a^2 + b^2 : a^2 - b^2$, where b is the radius of the free part of the lid, provided that no part of the base of the cylinder be left free from the fluid.

16. A circular tube of small section is half full of liquid and in the surface at each side floats one of two small equal spheres which just fit the tube. The tube rotates with angular velocity ω about a vertical diameter: find the pressure of the liquid at any point and shew that for values of ω greater than a certain quantity the pressure is a maximum at a depth $g\omega^{-2}$ below the centre of the tube.

17. A hollow cylinder is filled with water and made to revolve about a vertical axis attached to the centre of its upper plane face with a velocity sufficient to retain it at the same inclination to the axis. Find at what point of the plane face a hole might be bored without loss of fluid.

18. A vertical cylinder, height h and radius a containing water rotates with uniform angular velocity about its axis; if $1/n$th of the axis of the cylinder was above the surface before the rotation was established, prove that the greatest angular velocity, consistent with no escape of the water, is

$$\frac{2}{a}\sqrt{\frac{gh}{n}}. \qquad See \ 11$$

19. An open hemispherical cup, filled with water, is placed on a horizontal table, and the whole is made to rotate uniformly about its vertical radius; prove that the pressure on the table : the original weight of liquid as $8g - 3\omega^2 r : 8g$.

20. A hollow vessel in the shape of a wedge of a cylinder, formed by two planes through its axis, is filled with water and closed at the top; it is then made to rotate uniformly about the axis, which is vertical; find the pressure on the top, and the whole pressure on the curved surface of the cylinder.

21. A weightless cone is very nearly filled with liquid and inverted on a horizontal table; the liquid is made to rotate with an angular velocity ω, and the pressure required to keep the cone in contact with the table is equal to three times the weight of the liquid; prove that

$$3hw^2 = 4g \cot^2 a,$$

where h is the height of the cone, and a the semivertical angle.

22. Assuming that a mass of liquid contained in a vertical cylinder can rotate about the axis of the cylinder, under the action of gravity only, in such a manner that the velocity at any point varies inversely as the angular velocity of its distance from the axis of the cylinder; find the form and position of the free surface.

CHAPTER XV.

THE MOTION OF FLUIDS.

207. IF an aperture be made in the base or the side of a vessel containing liquid, it immediately flows out with a velocity which is greater the greater the distance of the aperture below the surface. The relation between the velocity and the depth, taking the aperture to be small, was discovered experimentally by Torricelli.

The following is Torricelli's Theorem :

If a small aperture be made in a vessel containing liquid, the velocity with which the particles of fluid issue from the vessel, into vacuum, is the same as if they had fallen from the level of the surface to the level of the aperture;

that is, if x be the depth of the aperture below the surface, and v the velocity of the issuing particles,

$$v^2 = 2gx.$$

The experimental proof of this is that if the aperture be turned upwards, as in the figure, the particles of liquid will rise to the same level as the surface of the liquid in the vessel. Practically the resistance of the air and friction in the conducting-tube destroy a portion of this velocity, but experiments tend to prove the truth of the law, which moreover can be established as an approximate result of mathematical reasoning.

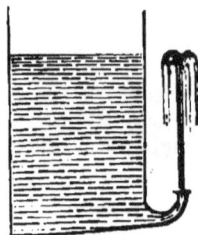

Assuming the principle of energy*, we can give a theoretical proof of the theorem.

Let K be the area of the upper surface of the liquid, and suppose that, during a short time, the height of the upper surface is diminished by a small quantity y, so that Ky is the volume which has flowed out through the orifice. Taking v as the velocity of efflux, the kinetic energy which has issued through the orifice is $\frac{1}{2}\rho Kyv^2$, and if we neglect the kinetic energy of the liquid in the vessel, this must be equal to the loss of potential energy, which is $g\rho Kyx$,

$$\text{and} \therefore v^2 = 2gx.$$

Form of a jet of liquid. If the aperture be opened in any direction not vertical, each particle of liquid having the same velocity, will follow the same path, which by the laws of Dynamics, is a parabola. Hence the form of the jet is a parabola.

Contracted vein. If the aperture be made in the base of a vessel, and if the base be of thin material, it is observed that the issuing jet is not cylindrical, but that it contracts for a short distance (a fraction of an inch) and then expands afterwards contracting gradually as it descends, and finally breaking into separate drops. The amount of contraction depends on the thickness of the vessel, and the size and form of the aperture.

The rate of Efflux is the rate at which the liquid flows out, and this clearly depends both on the velocity of the issuing particles, and the size of the aperture.

If k be the area of the aperture and v the velocity, then in an unit of time a portion of liquid will have passed through equal to a length v of a cylinder of which k is the base, and therefore vk is 'the quantity which flows out in an unit of time, that is, vk is the rate of efflux.

This is however not true unless the liquid issue from a pipe of some length, in which case there is no contracted vein. In general k must be taken as the section of the contracted vein, it being found that the velocity at the

* See Maxwell's *Matter and Motion.*

contracted vein is that which is given by Torricelli's
theorem.

208. *Steady motion.* When a fluid moves in such a
manner that, at any given point, the velocities of the successive particles which pass the point are always the same,
the motion is said to be steady. Thus if a vessel having a
small aperture in its base be kept constantly full, the motion
is steady.

Motion through tubes of different size. The continuity of
a fluid leads to a simple relation between the
velocities of transit through successive tubes.
Thus if a liquid, after passing through a tube
AB, pass through CD, the tubes being full, it
is clear that during any given time the quantity which passes a given plane AB in one
tube must be equal to the quantity which
passes any given plane CD in the other. Let
κ, κ' be the areas of these planes, and v, v'
the respective velocities at AB and CD. Then
κv, $\kappa' v'$ are the quantities which pass through
in an unit of time, and therefore

$$\kappa v = \kappa' v'.$$

Hence, as the section of a mass of fluid decreases, its
velocity increases in the same proportion. For instance,
the stream of a river is more rapid at places where the
width of the river is diminished. This also accounts for
the gradual contraction of the descending jet of liquid,
Art. (207), for the velocity increases, and therefore the
section diminishes.

209. *A cylindrical vessel containing liquid has a small
orifice in its base; to find the velocity at the surface.*

If the orifice be small and the surface large, the surface
will descend very slowly.

Let h be the height of the surface, then $\sqrt{2gh}$ is approximately the velocity at the orifice. Take K for the area of
the base of the vessel, and κ of the orifice.

Then, neglecting the change of velocity at the orifice in

the unit of time, $\sqrt{2gh}$ is the quantity of liquid which passes through the orifice, and therefore if V be the velocity at the surface,

$$VK = \kappa\sqrt{2gh}.$$

If the vessel be kept constantly full, the motion is steady and the velocities are constant: hence the time in which a quantity of liquid, equal in volume to the cylinder, would, under these circumstances, flow through the orifice

$$= \frac{h}{V} = \frac{K}{\kappa}\sqrt{\frac{h}{2g}}.$$

It will be seen that this is only a rough approximation to the actual facts of the case, but its insertion will serve to illustrate the laws above mentioned.

210. *Pressure of air in motion.* Early in the 18th century Hawksbee observed that if a current of air be transmitted through a small box the air becomes rarefied. This fact is illustrated by the following experiment.

Take a small straight tube, and at one end of it fix three smooth wires parallel to the tube and projecting from its edge, and let a flat disc be moveable on these wires, with its plane perpendicular to the axis of the tube. Blow steadily into the other end, and it will be found that the disc will not be blown off, but will oscillate about a point at a short distance from the end of the tube.

The reason of this apparent paradox is that the diminution of the density of the air in motion diminishes the pressure on the disc which would otherwise result from the continued action of the air impinging upon it, and the result is that it is balanced by the atmospheric pressure on the other side.

A full account of this experiment, and of other facts connected with it, was given by Professor Willis in the third volume of the *Cambridge Philosophical Transactions.*

A similar experiment was performed, in 1826, by M. Hachette, with a stream of water, and it was found that the pressure of the water was diminished by an increase of velocity.

It is worthy of remark that large ships of the Devastation class are observed to sink more deeply in the water when their speed is increased.

In a series of papers in *Nature*, for November and December 1875, Mr Froude has given a number of experimental illustrations, with explanations in general terms, of the connection between the pressure and the kinetic energy of a liquid.

Liquid in moving vessels.

211. *A cylindrical vessel, containing liquid, is raised upwards with a given acceleration; it is required to determine the pressure at any point of the liquid.*

Consider the motion of a thin vertical prism PQ of the liquid, having its upper end P in the surface, and observe that its vertical acceleration is caused by the pressure of the liquid on the end Q, the atmospheric pressure on the end P, and the weight of the prism.

If $PQ = z$, $p =$ the pressure at Q, $\kappa =$ the area of a horizontal section of the prism, and $f =$ the given acceleration, we obtain, by aid of the second law of motion,

$$\rho z \kappa f = p\kappa - \Pi \kappa - \rho z \kappa g;$$
$$\therefore \ p = \Pi + \rho z \,(g + f),$$

in poundals per square foot, or in dynes per square centimetre, according as the British or French absolute unit of force is employed.

212. *A box containing liquid slides down a smooth inclined plane; when the liquid is in a state of relative equilibrium it is required to find the pressure at any point and the surfaces of equal pressure.*

Every element of the liquid moves in a straight line with the constant acceleration $g \sin \alpha$, and, since the forces on any element are the resultant fluid pressure upon it and its weight, it follows that the resultant of these forces is $mg \sin \alpha$, parallel to the plane, m being the mass of the element.

It is hence easy to see that the resultant fluid pressure upon the element m is perpendicular to the plane and is equal to $mg \cos \alpha$.

Whatever be the shape of the element, the resultant fluid pressure upon it in the direction parallel to the plane is zero, and therefore it follows that the surfaces of equal pressure are planes parallel to the inclined plane, and that the surface of the liquid is a plane parallel to the inclined plane.

Taking z as the depth of a point in the liquid below the surface thus defined, and drawing a thin cylinder or prism from the point to the surface, the liquid enclosed will have no acceleration perpendicular to the inclined plane, and therefore, by the second law of motion, the pressure on the base of the prism will counterbalance the resolved part of the weight of the prism and the atmospheric pressure on its upper surface.

Hence if p is the pressure at the depth z and κ the area of the cross section of the prism,

$$p\kappa = \Pi\kappa + g\rho z\kappa \cos\alpha$$

or $$p = \Pi + g\rho z \cos\alpha.$$

We can also determine the surfaces of equal pressure when a box containing liquid is dragged up an inclined plane with a constant acceleration.

Taking f for the acceleration, and assuming that the liquid is in a state of relative equilibrium, it follows that the resultant pressure of the liquid on an element m, and the weight mg of the element have for their resultant the force mf parallel to the plane in the upward direction.

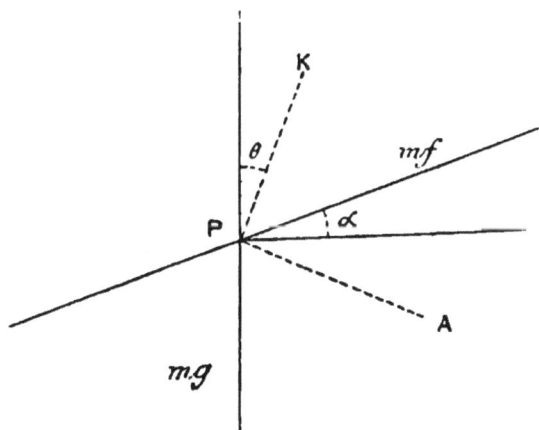

If then θ represent the inclination to the vertical of the direction PK of the resultant fluid pressure on an element, and if PA is perpendicular to PK, it follows, by resolving the resultant force and the acting forces in the direction PA, that

$$mf \cos(\alpha + \theta) = mg \sin \theta,$$

and $\therefore$ 　　　　$\tan \theta \{g + f \sin \alpha\} = f \cos \alpha.$

213.　*A closed vessel, in the form of a circular cylinder, is just filled with liquid, and the whole system rotates, as if rigid, uniformly about the axis of the cylinder; neglecting the action of gravity it is required to find the pressure at any distance from the axis.*

It is evident, if the cylinder is only just filled, that there is no pressure at any point of the axis. Taking any point P in the liquid, let PN be the perpendicular upon the axis, and consider the motion of a very thin column of liquid enclosing NP.

The pressures on the two ends of any small element of mass m of this column are such that their difference is equal to $m\omega^2 r$ in the direction PN, r being the distance of m from the axis.

The sum of all these differences is therefore equal to the sum of the values of $m\omega^2 r$ from N to P, and this sum, by Leibnitz's theorem, is equal to

$$\rho\kappa \, . \, NP \, . \, \omega^2 \tfrac{1}{2} NP.$$

But the sum of all the differences is equal to the pressure on the end P, *i.e.* is equal to $p\kappa$ if p is the pressure at P.

Hence we find that $p = \tfrac{1}{2}\rho\omega^2 NP^2.$

In the same manner if a closed vessel of any shape be just filled with liquid, and if the whole system be made to rotate as if rigid, about any fixed axis, which is entirely outside the vessel, it can be shewn that the pressure at any distance r from the axis is equal to

$$\tfrac{1}{2}\rho\omega^2 (r^2 - a^2),$$

where a is the distance from the axis of the point on the inside of the vessel which is nearest to the axis.

Impulsive Action.

214. Imagine a closed vessel filled with liquid and having an aperture in its surface fitted with a piston. Let an impulse be applied to this piston; then assuming the incompressibility of the liquid, it can be shewn by the same reasoning as for finite pressures, that the impulse is transmitted throughout the mass, and is, at any point, the same in every direction.

The impulse at any point is measured in the same manner as a finite pressure; that is, if ϖ be the impulsive pressure at a point, $\varpi\kappa$ is the impulse on a small area κ containing the point.

A cylindrical vessel, containing liquid, is descending with a given velocity and is suddenly stopped; to find the impulsive action at any point.

The impulsive pressure at all points of the same horizontal plane will be the same, and if ϖ be the pressure at a depth x, and κ the area of the base of the cylinder, $\varpi\kappa$ is the impulse between the portion of the liquid above and below the plane at a depth x, and this impulse evidently destroys, and is therefore equal to, the momentum of the liquid mass above, which is $\rho\kappa x v$.

Hence $$\varpi\kappa = \rho\kappa x v,$$

and $\therefore$ $\varpi = \rho v x$.

215. If a vessel of any shape, containing liquid, descend vertically and be suddenly stopped, we can prove, by considering a small vertical prism of liquid, that the impulse at any point varies as the depth below the surface of the liquid.

This being the case, it follows that the propositions relating to whole pressure, and to resultant vertical and horizontal pressures in Chapters III. and IV., are equally true of impulsive pressures for the particular case in which the motion destroyed is vertical. The question is really the same if the vessel be made to ascend suddenly from rest, or have its velocity suddenly changed.

Thus, if S is the area of any plane in the descending vessel in contact with the liquid, and $\bar{z}$ the depth of its centroid below the free surface of the liquid, the resultant impulse on the plane is $\rho v \bar{z} S$.

Also for any surface the expression for the whole impulse is $\rho v \bar{z} S$.

Again the resultant vertical impulse on any surface will be equal to the momentum of the mass of liquid contained between vertical lines through the boundary of the surface and the free surface of the liquid.

And further, the resultant horizontal impulse in a given direction, on any curved surface, will be equal to the resultant impulse on the projection of the given curved surface on a vertical plane perpendicular to the given direction.

If a closed vessel, just filled with liquid, be moved in any direction and suddenly stopped, the surfaces of equal impulse will be planes perpendicular to the given direction, and the free surface will be that plane which passes through the extreme particles of the liquid in the rear of the motion.

If z is the distance of any point in the liquid from this plane, the impulse at the point will be $\rho v z$, and all the theorems above given are equally true of this case.

Ex. 1. *A hollow sphere just filled with liquid is let fall upon a horizontal plane.*

The resultant impulse, downwards, on the lower half of the internal surface

$$= \rho v \left\{ \pi r^3 + \frac{2}{3} \pi r^3 \right\} = \frac{5}{3} \rho v \pi r^3,$$

and the impulse, upwards, on the upper half

$$= \rho v \left\{ \pi r^3 - \frac{2}{3} \pi r^3 \right\} = \frac{1}{3} \rho v \pi r^3.$$

The resultant vertical impulse on either of the halves made by a vertical plane $= \frac{2}{3} \rho v \pi r^3$, and the resultant horizontal impulse $= \rho v \pi r^3$.

Ex. 2. *In a closed vessel of liquid a ball of metal is suspended by a vertical string fastened to the upper part of the vessel. Find the impulsive tension of the string when the vessel is suddenly raised with a given velocity.*

The resultant impulse of the liquid on the ball will be the same as if its place were occupied by the liquid, and therefore will be equal to the momentum of the ball of liquid.

If U be the volume, and v the velocity, this is $\rho v U$. But if ρ' be the density of the metal, the momentum of the ball is $\rho' v U$, and this is produced by the impulse of the liquid, and the tension T of the string.

Hence $\qquad \rho' v U = \rho v U + T,$

and $\qquad T = (\rho' - \rho) v U.$

216. *Pressure produced by a jet of liquid impinging on a plane.*

Imagine a vertical jet of water to fall with a given velocity on a horizontal plane. If we assume that there is no splashing and that the water flows away on the plane, the vertical momentum of the jet will be destroyed on reaching the plane.

Hence if ρ is the density of the water, v the velocity, and κ the cross section of the jet when it impinges on the horizontal plane, the momentum destroyed in the unit of time is $\rho\kappa v \cdot v$ or $\rho\kappa v^2$.

This then being the rate at which momentum is being destroyed is the measure of the pressure in the plane.

Taking a foot and a second as units of length and time, the pressure is given in poundals.

If the jet fall on an inclined plane, the momentum perpendicular to the plane is destroyed.

Now the amount destroyed in the unit of time is

$$\rho\kappa v \cdot v \cos \alpha.$$

The pressure is therefore $\rho\kappa v^2 \cos \alpha.$

EXAMPLES.

1. A hollow cone, whose vertical angle is given, is filled with water and placed with its base on a horizontal plane; determine a point in its surface at which, an orifice being made, the issuing fluid will just fall outside the base of the cone.

2. Through the plane vertical side of a vessel containing fluid, small holes are bored in the circumference of a circle, which has its highest point in the surface of the fluid; shew that the trace of the issuing fluid on a horizontal plane through the lowest point of the circle is two straight lines.

3. Two cylindrical vessels containing water are suspended with their axes vertical to the ends of a string passing over a fixed smooth pulley in a vertical plane; neglecting the weights of the vessels, compare the whole pressures, during the motion, on the curved surfaces of the cylinders.

4. A hollow cone, vertex downwards, and containing liquid, is attached to a string passing over a pulley and supporting at its other end a given weight: determine the motion and find the whole pressure of the fluid on the cone and also the resultant pressure.

5. Two hollow cones, filled with water, are connected together by a string attached to their vertices which passes over a fixed pulley; prove that, during the motion, if the weights of the cones be neglected, the total pressures on their bases will be always equal, whatever be the forms and dimensions of the cones. If the heights of the cones be h, h', and heights mh, nh' be unoccupied by water, the total normal pressures on the bases during the motion will always be in the ratio

$$n^2+n+1 : m^2+m+1.$$

6. Two spherical shells of the same material and of thicknesses proportional to their radii are each half filled with water. Prove that when tied to the ends of a string slung over a smooth pulley, and allowed to move, the resultant pressures of the water on the spheres are equal.

7. A ball of lead is let fall in water; assuming that the pressure of the water is the same as if the ball were not in motion, find its velocity at any given depth.

8. An Hydraulic Ram being in action, V, v are the mean velocities of the falling and rising water, and $n : 1$ is the ratio of the sectional areas of the pipes; prove that the height to which the water is raised is $nV^3/2gv$, neglecting the waste of energy by friction or overflow.

9. When water is flowing along a pipe, the friction varies as the square of the velocity and the surface of the pipe. If a velocity of 12 feet per second causes a friction of 1 lb. per square foot of wetted surface, what will be the friction on 1 mile of pipe 7 inches in diameter, the water flowing at the rate of 3 ft. per second? What will be the loss of horse-power in transmitting energy through this pipe at this rate?

10. If A is the area of the section of each pump of a fire engine, l the length of the stroke, n the number of strokes per minute, and B the area of section of the hose, find the mean velocity with which the water rushes out.

11. A box containing water is projected up a rough inclined plane, the inclination of which to the horizontal is greater than the angle of friction; shew that the free surfaces of the fluid in its position of rest relative to the box when going up and coming down are planes inclined to one another at an angle equal to twice the angle of friction for the box and the inclined plane.

12. A railway train, travelling with a given acceleration, arrives at an incline, and, after ascending to a ridge, descends at the same incline on the other side. Assuming that the pull of the engine and the resistance are the same throughout, determine the levels of the water surface in the boiler in going up and down the incline, and prove that the difference of the levels is equal to the angle between the inclines.

13. Two smooth inclined planes are fixed back to back, and two boxes containing liquid slide on the planes under gravity, the boxes being connected by a fine string passing over a pulley at the vertex of

the planes. Prove that the free surfaces of the liquids will be parallel and equally inclined to the planes, if the weights of the boxes and the liquids they contain are proportional to the cosecants of the angles which the inclined planes make with the vertical.

14. A circular tube of fine bore, whose plane is vertical, contains a quantity of heavy uniform fluid, which subtends an angle $2a$ at the centre; a heavy spherical particle, just fitting the tube, is let fall from the extremity of a horizontal radius; find the impulsive pressure at any point of the fluid.

15. A cylindrical vessel containing inelastic fluid is descending with a given velocity (v) and is suddenly stopped; its axis being vertical, find the whole impulse on the curved surface.

16. A hollow cone, formed of flexible material, is filled with water, and closed by a rigid circular plate. It is then let fall, with its axis vertical and vertex upwards, on a horizontal plane; find the whole impulse, and the resultant impulse, on the curved surface.

Determine also the impulsive tension, at any point, in the direction of the generating line through the point.

17. A closed vessel is filled with water containing in it a piece of cork which is free to move; if the vessel be suddenly moved forwards by a blow, prove that the cork will shoot forwards relative to the water.

18. If the surface of the earth were a perfect sphere, prove that a river flowing uniformly due south, in latitude 45^0, would be going up an incline of, apparently 1 in 570, in consequence of the earth's rotation.

19. Prove that the longitudinal tension per unit of length of a flexible pipe of uniform bore, in the form of a circle, due to water flowing through it with constant velocity v, is ρv^2, where ρ is the mass of water per unit of length of the pipe; and hence prove that if the pipe be at rest in any curve under any forces, the equilibrium will not be disturbed if the water in the pipe be flowing with constant velocity, and that the tension at every point will be increased by ρv^2.

20. A bucket filled with water to the depth of a foot is suspended from a balance. A small aperture is opened in its base which discharges 4 lbs. of water per minute at an angle of 45^0 with the vertical, and simultaneously a stream of water is received by it which emerges vertically from an aperture 8 feet above the free surface with a velocity of 30 feet per second, and falls on a glass plate attached to the bucket obliquely just above the free surface so that splashing is prevented, the supply being such that the level in the bucket remains constant. Prove that the balance indicates about ·066 lb. more than the weight of the bucket and its contents.

CHAPTER XVI.

216 *a*. THE sensation which we call sound is produced by a vibratory movement of the atmosphere; however it is first caused, it finally affects the organs of hearing by means of the air. A blow struck on any elastic body will produce sound, and the more highly elastic the body is the more easily will the sound be produced; a piece of metal when struck will ring sharply while the same blow on a piece of wood produces a dull sound of less intensity. A sound may traverse intervening bodies and be finally imparted through air which has no direct communication with the air in which it originated.

The fact that *air is necessary for the transmission of sound* may be shewn experimentally. Suspend a bell within the receiver of an air-pump, and provide a means of striking the bell from without, for instance, by a rod sliding in an air-tight collar. Then proceed to exhaust the receiver, and it will be found that as the exhaustion progresses, the sound of the bell becomes fainter, and is finally lost altogether.

That there is an actual motion in the particles of air is shewn by the transmission of sound through solid bodies, and also by the fact that a musical note sounded on any instrument will sometimes produce a sound, in unison with it, from some other body not in contact with the instrument.

Velocity of Sound.

The rate at which sound travels depends on the temperature of the atmosphere; it has been found experimentally that at the freezing temperature the velocity is about 1089 feet per second, and that at a temperature of 61° F., when the height of the barometer is 29·8 inches, the velocity is nearly 1118 feet per second. We may therefore take 1100 feet per second as the velocity of sound under average atmospheric conditions.

Distance of a sounding body. Knowing the velocity of sound, we can estimate the distance of a sounding body whenever the production of sound is accompanied by the production of light. The velocity of light is so great that its transmission through all ordinary distances on the earth may be considered instantaneous, and thus if a cannon be fired from a ship at sea, the interval between seeing the flash and hearing the report will determine the distance of the ship. In the same manner the interval between a flash of lightning and the thunder which follows it will determine the distance of the cloud from which the flash is evolved.

The rolling of thunder may be accounted for in two ways. A single explosion may accompany the lightning, in which case a peal of thunder will be due to the reflection of the sound by clouds in different directions, and will be in fact a succession of echoes. Or the electric flash may pass rapidly from cloud to cloud, and thus the sounds of a series of explosions taking place almost at the same instant, but at different distances from the spectator, will arrive in succession and produce a continuous peal. In this latter case the peal is probably intensified and lengthened by echoes.

Velocity of sound through water. Sound is transmitted with much greater velocity through water, and through highly elastic solids, than through air. By experiments made in the lake of Geneva, the velocity was found to be 4708 feet per second, when the temperature of the water was 8° C. The rate of transmission through metallic substances is very much greater.

Velocity through gases. We have stated that the velocity in air depends on the temperature, and not on the density. In fact it depends on the value of k, which is different for different gases, and therefore the velocities in gases differ from each other. For instance, the velocity in hydrogen is nearly four times that in air at the same temperature, the elasticity of hydrogen being much greater than that of air.

Transmission through the atmosphere. The various portions of the atmosphere through which a sound passes, may have different temperatures, and consequently the sound will travel with a variable velocity. Moreover, the passage through varying strata tends to disturb the vibrations and to diminish the intensity. This accounts for the fact that distant sounds are heard more distinctly at night than during the day, the atmosphere being in general more quiescent, and having a more equable temperature.

Sound Waves.

216 b. A *wave* is the term applied to any state of vibratory motion which is transmitted progressively through the particles of a body. The effect of dropping a stone in still water is a familiar illustration; the rise and fall of the water produced by the plunge of the stone travels outwards in an expanding circle, while the particles of water merely rise and fall in succession as the wave passes over them.

Thus a portion of the atmosphere being in any way set in motion, the vibrations are communicated to the surrounding air, and the expanding spherical wave impinging on the ear produces the sensation of sound.

The intensity of a sound diminishes as the distance of the sounding body is increased. As a spherical wave expands, its thickness remaining constant, the vibrations are communicated to larger masses of air, and, in accordance with a general law of mechanics, the intensities of the vibrations are diminished. The intensity in fact is diminished in the inverse ratio of the square of the distance. This law however does not hold, if the sound be transmitted through tubes or pipes. In such cases the intensity is very slowly diminished.

Propagation of a Wave along a Straight Tube.

217. Consider a straight tube filled with air, and let a disc AB at one end oscillate rapidly over the space aa'.

When the disc oscillates from A to a, it compresses the air before it, and when the disc is at a, the compression has traversed and extends over the space AC. This compression travels along the tube with a constant velocity, and is called the condensing wave.

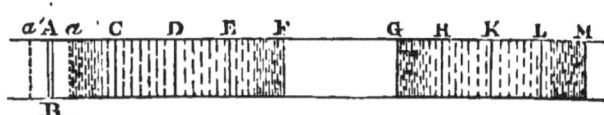

As the disc returns from a to A, it rarefies the air behind it; and this rarefaction extends over AC, while the previous compression has been transferred to the space CD, and thus a rarefying wave follows the condensing wave.

As the disc moves from A to a', another rarefying wave is produced, and when the disc returns to A, a condensing wave is produced, while during these two processes the first condensing and rarefying waves have been transferred to EF and DE respectively.

The disc having its greatest velocity at A, and coming to rest at a and a', it is obvious that the condensation is greatest at F, and diminishes gradually to E, where there is no condensation, or where the density is the same as if the air were at rest; from E to D the air is rarefied, and at D the rarefaction is greatest; from D to C the rarefaction decreases, and at C condensation commences and increases to A.

Thus a complete wave or undulation is formed, and if the disc oscillate once only, a single wave will travel along the tube taking successive positions as in the figure; if the disc continue to vibrate, a succession of these waves will be produced and will follow each other continuously along the tube. If these waves, on emergence from the tube, impinge on the ear, the sensation produced will be that of a continuous and uniform sound.

The vibrations can be produced without the aid of the disc, as, for instance, by blowing across the end of the tube.

It will be observed that the velocities of the vibrating particles of air are zero at F and D, and greatest at E and C.

The length of a wave is the distance between any two points at which the *phases* of vibration are the same, that is, at which the velocities of the vibrating particles are the same in direction and magnitude.

Motion of a Wave along a Stretched String.

218. In a similar manner, if a portion of a stretched cord *PQ* be set in motion, a wave, or succession of waves, will travel along the cord, and on arriving at *Q* will be reflected and travel back again.

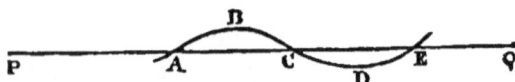

The string may vibrate somewhat in the form of the curve *ABCDE*, *AE* being the length of a wave, *B* and *D* the points at which the displacement is greatest, and the velocity zero, and *A*, *C*, and *E* the points at which the displacement is zero and the velocity greatest.

In this case the vibration is perpendicular to the line in which the wave travels, but its analogy with the case of the tube is sufficiently evident.

The vibrations of the string are communicated to the air and thereby conveyed to the ear.

Musical Sounds.

219. Any series of waves, following in close succession, may produce a continued sound; if they are irregular in magnitude, the result is a *noise*, but a *musical note* is produced by a constant succession of equal waves.

Pitch, intensity, and quality. Notes may differ from each other in three characteristics; thus, a note may be grave or acute, that is, its *pitch* may be high or low; and the pitch of a note depends on the length of the constituent wave, and is higher as the length of the wave is less. The *intensity* of a sound depends on the extent of vibration of the particles of air, and its *quality* is a characteristic by which notes of the same pitch and intensity are distinguished from each other. The quality of a note, or, as it is sometimes called, its *timbre*, depends on the nature of the instrument from which it is produced.

A further distinction of sounds is sometimes marked by the word *tone*. Thus the tone of a flute differs from the tone of other instruments, while two flutes may, and will in general, produce sounds which differ in quality.

Sounds of different pitch travel with the same velocity. This appears to be the case from the fact that if a musical band be heard at a distance there is no loss of harmony, and therefore there can be no sensible difference in the velocities of the different sounds.

220. *Reflection of waves in a tube of finite length.* It is found both by experiment and theory that a wave on arriving at the end of a tube is reflected, whether the end be open or closed, and travels back again, changed only in intensity, to be again reflected at the other end.

This accounts for the *resonance* in a tube when the air within it has been set in vibration.

221. *Coexistence and interference of undulations.*

Different sound waves travelling through the air traverse each other without alteration either of pitch or intensity. In other words, different undulations coexist without affecting each other, and the actual vibration of any particle of air is the sum or difference of the coexistent vibrations which are at the same instant traversing the particle of air.

A simple illustration of this coexistence may be seen by dropping two stones in water. The expanding circular waves intersect, and at the points of intersection it will be seen that a depression and an elevation neutralize each other, and that two depressions or two elevations at the same point increase the amount of one or the other. If there be a sufficient number of circular waves the points of greatest elevation will be seen to lie in regular curves, as also those of depression, and of neutralization*. The vibrations in this case being transverse to the direction of transmission of the wave are different from those of sound waves, which are longitudinal or in the direction of transmission, but the effect of coexistence is the same in all cases.

The effect of coexistence in producing neutralization, or increase of intensity, is called the interference of undulations, and it must be observed that, while two sets of undulations are physically independent of each other, their geometrical resultant may be a form of undulation different from that of either component, as in the case just referred to of the undulations in the surface of water.

The Notes which can be produced from a Tube closed at one end.

222. When a definite note is being sounded from a tube, the air within the tube vibrates regularly, every particle maintaining the same vibration, and there are certain points of the tube at which the air remains at rest. These points, or planes of division of the tube, are called *nodes*, and the planes of maximum vibration are called *loops*.

The motion in fact is the same as if there were fixed waves in the tube, and the nodes and loops are the points of zero velocity and zero condensation.

The motion thus described is called *steady* motion, and its existence is requisite to the continuance of a definite note.

In the case of a tube closed at one end B, it is clear that the end B must be a node, and since the end A is open its

* These curves are hyperbolas, for, if A and B be the centres of disturbance, and P, P' the points of intersection of two particular waves, AP and BP increase uniformly with the time, and the rate of increase of each is the same.

Hence, their difference is constant, and the locus of P is an hyperbola of which A and B are the foci. As other waves follow in succession the series of such points will lie in confocal hyperbolas.

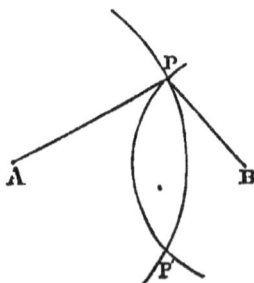

density is sensibly that of the air outside, and we may take it to be a loop.

It is therefore evident that the longest possible wave for which the motion can be steady is four times the length of AB; and the corresponding sound is the fundamental note of the tube.

Further, AB may be any odd multiple of the distance from a node to a loop, and if $AB = l$, and λ be the length of a wave, we must have

$$l = (2n+1)\frac{\lambda}{4},$$

$$\text{or } \lambda = \frac{4l}{2n+1}.$$

Hence the notes which can be produced from AB have for their wave lengths,

$$4l, \ \frac{4l}{3}, \ \frac{4l}{5}, \ \&c.,$$

and, if v be the velocity with which a wave traverses the tube, the times of vibration are

$$\frac{4l}{v}, \ \frac{4l}{3v}, \ \frac{4l}{5v}, \cdots$$

and are therefore in the ratio of the fractions

$$1, \ \frac{1}{3}, \ \frac{1}{5}, \ \frac{1}{7}, \ \&c.$$

The Notes of a Tube open at both ends.

223. In this case each end is a loop, and there is therefore a node between; hence the greatest possible wave length is twice the length of the tube, and further the length of the tube must be some multiple of half the length of a wave.

Hence $\qquad l = m\frac{\lambda}{2}$, and $\lambda = \frac{2l}{m}$.

The successive wave lengths are therefore

$$2l, \ l, \ \frac{2l}{3}, \ \frac{l}{2}, \ \frac{2l}{5}, \ \&c.$$

and the vibrations in the ratio of the fractions

$$1, \frac{1}{2}, \frac{1}{3}, \frac{1}{4}, \dots$$

It will be observed that the fundamental note of the open tube is an octave higher than that of the closed tube of the same length, the wave length for the former being half that for the latter.

The Formation of Nodes and Loops.

224.　Taking the case of a tube HK closed at the end K, the aerial particles at the end K are permanently at rest, while those at A are in a state of permanent vibration.　We have stated, as an experimental fact, that a series of waves travelling along HK in regular succession are reflected at K and travel in the opposite direction; and this fact enables us to account for the existence of nodes and loops.

In order to give the required explanation we must first explain a method of representing geometrically the state of motion of the aerial particles in a wave.

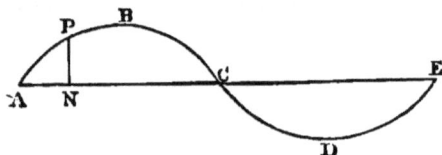

Take AE as a wave length, and let the ordinates of the curve $ABCDE$ represent the velocities of the several particles parallel to the line AE; thus NP represents the velocity of the particle at N, NP being drawn upwards when the velocity is in the direction AE, and downwards when the velocity is in the opposite direction.

Hence, if two distinct sets of vibrations coexist along a line of aerial particles, we can determine the resultant motion by drawing the two curves for the two waves, and the algebraic sum of the ordinates at any point will represent the resultant velocity at that point.

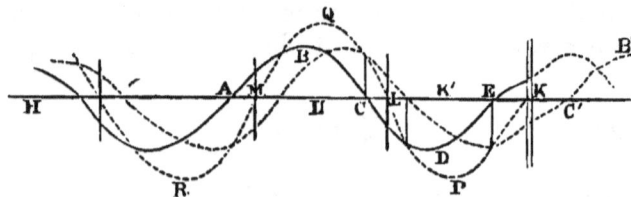

Imagine now a wave travelling along AB, and impinging on the fixed end K; this wave will be reflected and will travel along BA with reversed velocity.

In the figure the dotted line $B'C'$ will represent the reflected wave, and the effect of the reflected wave is the same as that of a wave $B'C'$ travelling in the direction KA.

It will be seen from the figure that the velocities at K from the two waves are always equal and opposite, and that the resultant velocity at K is always zero, in accordance with the given conditions. In other words, the effect of the fixed end K is replaced by the effect of a reversed wave travelling in the opposite direction.

It will also be seen that there is a succession of points, K, L, M,... at which the velocity is always zero and a succession K', L',... at which the velocity varies between its greatest values in both directions, the former set of points being nodes, and the latter loops.

Let a dotted line $KPLQMR$ be drawn such that its ordinate at any point is the algebraic sum of the ordinates corresponding to the incident and reflected wave ; this dotted line will represent the state of vibration of the air in the tube at the instant considered, and it will be observed that while the points K, L, M,... are points of permanent rest, all the intermediate points represent the positions of aerial particles which vibrate steadily, their velocities being zero at regular intervals.

Thus, the opposing waves may be so placed that their extremities C, C' may coincide at K''; in the figure this will occur when the incident waves have traversed the space CK'', and the opposing waves the space CK'', and at this instant the velocity at K' will be zero. Subsequently the two waves travelling in opposite directions will produce at K' a velocity double that of either, so that the velocity at K' will then be a maximum, the interval of time being that during which the vibration has traversed a space equal to one-fourth of a wave length.

It will be now clear that, if a permanent vibration be maintained at the open end H, a succession of nodes and loops will necessarily be formed in the tube, provided that the wave length emitted from the end H is such as to satisfy the condition of Art. (222). This condition is that the wave length should be an odd submultiple of 4 times the length of the tube.

In a similar manner, if KH be a tube open at both ends, it is found that a wave or a set of waves travelling along HK are reflected at K, and traverse the tube in the direction HK. An important difference however exists between the two cases ; in the former case the end K is a node, in the present case it is a loop, the particles of air vibrating freely, and the density being the same (very nearly) as that of the external air.

An analogous explanation will account in this case also for the formation of nodes and loops.

In the case of the tube closed at K, the reflected wave on arriving at H is partly emitted into the open air and partly reflected, thereby reinforcing the new vibration which is at the instant being excited at H, and aiding to produce another series of waves which are again reflected at K.

The effect which is thus produced on the ear is that of a sustained note, the character of which depends on the material of which the tube is formed, while its pitch and intensity depend solely on the lengths of the constituent waves and the extent of the vibrations of the aerial particles.

The Notes produced by a Vibrating Cord.

225.   A stretched cord in a state of vibration may either oscillate as a whole, figure (1), or in parts, as in figures (2) and (3), the curved lines representing the actual positions at certain instants of the cord itself.

In any case the two ends are points of zero velocity or nodes, and the wave corresponding to the fundamental note has, for its length on the cord, twice the length of the cord.

In general,—if the wave length on the cord be λ', the length l of the cord must be some multiple of $\frac{1}{2}\lambda'$,

$$\text{i.e. } l = m\frac{\lambda'}{2}.$$

The velocity of propagation along the cord will depend on its tension, thickness and density; and if v' be this velocity the time of vibration is $\frac{\lambda'}{v'}$.

The pitch of the note produced is determined by the time of vibration, and therefore, if λ be the wave length produced in air by the vibrations of the cord and thereby conveyed to the ear as a sound, and v the velocity of propagation in air, we shall obtain the note by the relation

$$\frac{\lambda}{v} = \frac{\lambda'}{v'},$$

since the aerial vibrations are performed in the same time as those of the cord.

Hence
$$\frac{\lambda}{v} = \frac{2l}{mv'},$$

and the wave lengths are

$$2l\frac{v}{v'}, \quad l\frac{v}{v'}, \quad \frac{2l}{3}\frac{v}{v'}, \quad \frac{l}{2}\frac{v}{v'}, \quad \&c.$$

These wave lengths give the series of harmonics producible from the cord, and it should be observed that any one may be produced alone, or any number of them may exist simultaneously.

226. *Vibration of rods.* We know that sounds are produced by vibrating rods, and we can determine the series of notes producible in any simple case by the considerations of the preceding articles. A rod fixed at one end and free at the other, will have for its fundamental note a wave length four times its own length, the fixed end corresponding to a node and the free end to a loop.

The analogy between a vibrating rod and a vibrating column of air will be now seen, but attention must be paid to the fact that the vibrations of air which produce sound are longitudinal, while the vibrations of a string are transversal, and those of a rod may be either transversal or longitudinal.

A common instance occurs in the humming of a telegraph-post, which is probably due to a series of longitudinal vibrations traversing the post in a vertical direction.

The transmission of sound through water is analogous to the transmission of sound by means of longitudinal vibrations along a rod, and is treated theoretically in exactly the same manner.

227. *The pitch of a note* produced by a vibrating cord depends on the tension and substance of the cord, and is heightened by an increase of tension; and in a similar manner the pitch of a note produced from a rod is found to depend on its size and substance.

This is due to the fact that the rates of propagation of vibrations depend on the characteristics above mentioned, and thus a long wave length, traversing a cord or a rod very rapidly, may give rise to a short wave length in the aerial vibrations which result from those of the cord or the rod, and a high-pitched note be produced.

Unison and Harmony of Musical Notes.

228. The vibration number of a note is the number of vibrations imparted to the air in the course of one second.

If n is the vibration number of a note, τ the time of vibration, λ the wave length, and v the velocity of sound, we have the relations

$$n\tau = 1, \quad \lambda = v\tau.$$

Two notes are said to be in *unison* when the times of vibration, or the wave lengths, are the same for both.

The *harmony* of two notes consists in the recurrent coincidence, at short intervals, of their constituent vibrations; thus, if a note and its octave be sounded, the vibration belonging to the fundamental note coincides exactly with two vibrations of the octave, and the two sounds are said to be in harmony with each other.

More generally two notes are in harmony when a small number of vibrations belonging to one of them coincides exactly, in time, with a small number of the vibrations belonging to the other. An instance of this is the harmony of a note with its fifth in the diatonic scale, three vibrations of the upper note being coincident with two of the lower note.

229. *Communication of vibrations.* If two different bodies can vibrate in unison or in harmony with each other, that is, if their fundamental notes are either in unison or in harmony, it is a known fact that when one is set vibrating, the other, if not too far off, will vibrate also. The reason is that the sound waves diverging from one body impinge on the other, and when the vibrations of the latter can be in harmony with those of the former, the slight vibration at first established is maintained and intensified by the continued impulses of the same aerial vibration. Thus a person singing or whistling in a room may sometimes hear notes sounding from thin glass jars or metallic tubes, and these

notes will always be in harmony with the note originally sounded.

230. *Beats.* When a note of a pianoforte is sounded a series of alternations is generally to be noticed in the intensity of the sound, these alternations, which are called *beats*, occurring at regular intervals.

This phenomenon depends on the fact that there are in general two strings to each note, which are intended to be exactly in unison with each other. Practically the unison is seldom perfect, and hence the two sets of waves do not exactly coincide with each other.

The intensities are however very nearly the same, and hence, when the vibrations of the two waves oppose each other, a diminution of the intensity results, but when they are in the same direction the intensity is increased.

Suppose that τ and τ' are the times of vibration of the two notes; then if x vibrations of one coincide with $x + 1$ of the other, we have

$$\tau x = \tau' (x + 1),$$

or

$$x = \frac{\tau'}{\tau - \tau'};$$

and $\therefore \frac{\tau \tau'}{\tau - \tau'}$, is the interval between the instants of time at which the vibrations oppose each other, and is therefore the period of the beats.

It follows that the number of beats in one second

$$= \frac{\tau - \tau'}{\tau \tau'} = \frac{1}{\tau'} - \frac{1}{\tau},$$

which is the difference of the vibration numbers of the two notes.

It is evident that the more nearly τ and τ' are equal to each other, the longer is the period of the beats, and the less the number of beats heard while the sound is perceptible.

Beats are also produced when two notes are very nearly in harmony with each other; the explanation is the same as for the simple beats above mentioned.

Tartini's Beats. Again, when two notes are actually in concord, a note is sometimes heard in addition to the two notes, and of lower pitch than either. The vibrations of the two notes coincide at regular intervals; these coincidences are Tartini's beats, and the effect of a series of such beats, at regular and rapidly recurring intervals, is that of a note which is grave in comparison with the original notes. This lower note is called a *subharmonic* of the two notes by which it is produced.

For example, in the case of a perfect fifth every second vibration of one note coincides with every third of the other, and the effect produced is that of a note exactly one octave below the lowest note of the concord.

In general, if in a certain fraction (τ), of a second, one note makes m vibrations and the other n, the period of vibration of the resultant subharmonic is τ, m and n being supposed to be prime to each other.

231. *Effect of motion upon the apparent pitch of a note.*

It is well known as a matter of observation that, if an express train passes a station at high speed, a note sounded in the train appears to a listener at the station to be of higher pitch as the train approaches the station, and of lower pitch when the train is receding from the station.

To explain this fact we must bear in mind that the apparent pitch of a note, to a listener, depends upon the number of vibrations which enter his ear in one second.

Let n be the frequency, or vibration number, of the note sounded in the passing train, and let u be the velocity of the train per second, and v the velocity of sound per second.

Taking P as the position of the listener at the station, let A, A' be positions passed by the approaching train at the interval of one second, and B, B' positions passed by the receding train at the interval of one second, so that $AA' = BB' = u$.

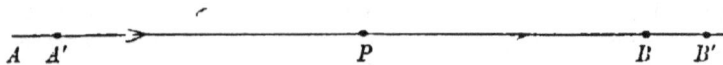

The time between the first and the nth succeeding vibration arising at P

$$= 1 + \frac{A'P}{v} - \frac{AP}{v} = 1 - \frac{u}{v}.$$

The listener then receives n vibrations in the time $\dfrac{v-u}{v}$,

and therefore he receives $\dfrac{nv}{v-u}$ vibrations in one second.

Hence the apparent frequency of the note is $\dfrac{nv}{v-u}$.

Again when the train has passed the station, the listener receives n vibrations in the time

$$1 + \frac{B'P}{v} - \frac{BP}{v} \text{ or } 1 + \frac{u}{v}.$$

The apparent frequency of the note heard is therefore $\dfrac{nv}{v+u}$.

Next take the case in which the note is sounded at the station and the listener is in the train.

Taking A for the station let PP' be the space passed over by the approaching train, and QQ' by the receding train while the listener in the train receives n vibrations.

Then if t is the time from P to P',

$$t = 1 + \frac{AP'}{v} - \frac{AP}{v} = 1 - \frac{ut}{v}, \text{ and } \therefore t = \frac{v}{v+u}.$$

The frequency of the apparent note is therefore $n\,\dfrac{v+u}{u}$.

Again, for the receding train, if τ is the time from Q to Q',

$$\tau = 1 + \frac{AQ'}{v} - \frac{AQ}{v} = 1 + \frac{u\tau}{v}, \text{ and } \therefore \tau = \frac{v}{v-u}.$$

The frequency of the apparent note is therefore $n\,\dfrac{v-u}{v}$.

For instance, if the train is going 60 miles an hour, $u=88$, and, if the note sounded is the middle C, the frequency of the apparent note to the listener in the train when approaching the station is $264 \times \dfrac{1178}{1090}$ or 285; that is, the note is raised rather more than a semitone.

NOTES.

Velocity of Sound.

A calculation from theoretical principles of this velocity was made by Newton and again by Lagrange ; the result obtained was about 916 feet per second.

This notable discrepancy between fact and theory remained unexplained until Laplace remarked that the heat developed by the sudden compression of the air would increase the elasticity, and therefore increase the calculated velocity.

New calculations were made, and the result is in complete accordance with fact.

If we neglect the heat developed by the sudden compression of air, the theoretical expression for the velocity of sound is $\sqrt{k}$, where k is the ratio of the numerical values of the pressure and the density of air, the pressure being measured in poundals per square foot, and the density in pounds. Since k is approximately 840000, i.e. g times the height of the homogeneous atmosphere, Art. (194), it will be seen that its square root gives very nearly the number above mentioned.

If we take into account the heat developed by compression, the theoretical expression is $\sqrt{k\gamma(1+at)}$, where γ is the coefficient introduced by the consideration of such development, and t is the temperature.

The numerical value of γ is about 1·414*.

Intensity of sound after traversing pipes. Experiments were made by Biot with some water-pipes in Paris, and it was found that a whispered conversation could be carried on through a pipe 3000 feet in length.

The use of speaking-tubes in large houses is another illustration of the fact mentioned in Art. (216).

Vibrating Cords. It is found that the velocity with which a wave traverses a stretched cord is the same as the velocity which would be acquired by a heavy body falling through a vertical space equal to half that length of the cord the weight of which is equal to its tension. In other words, if the weight of a length l of the cord be equal to its tension, the velocity with which a wave travels along it is $\sqrt{gl}$.

The existence of nodes and loops in the case of a cord may be practically manifested by placing on the vibrating cord small pieces of paper, cut so as to rest on the cord ; those which are placed at the

---

* See Lord Rayleigh's *Sound*, Chapter xi.

nodes will remain on the cord, while those which are placed near the loops will be thrown off.

The *Monochord* is a simple instrument for trying the experiments just mentioned, and for testing other results of theory.

A cord fastened at one end is stretched over a sounding board, and passing over a bridge is tightened by a weight at the other end ; the tension may be varied by changing the weight, and by means of another bridge, moveable along the board, the length of the vibrating portion may be diminished. The notes obtained for different lengths and different tensions can be thus compared, and the wave lengths for different notes can be directly measured.

Practical illustration of the interference of aerial vibrations. From Art. (221) we can see that if two waves, exactly similar to each other, travel in the same direction, and one be half a wave length behind or before the other, the result will be a permanent quiescence of the aerial particles along the direction in which the waves travel.

This has been shewn visibly by an experiment, which is due to Mr Hopkins.

A straight tube AB branches off at the end B into two portions BC, BD; the end A is closed by a tight membrane and fine sand is scattered over the membrane. A vibrating plate of glass is placed beneath C and D so that the two portions immediately beneath C and D shall be in opposite *phases* of vibration. The waves thus produced in CB and DB traverse these branches of the tube, and arrive at B in opposite phases, that is, one is the half of a wave length before the other, and therefore there is theoretically no resultant vibration in BA. Practically it is found that the sand on A is undisturbed, but, if the plate be turned round, the sand is immediately thrown into a state of violent commotion.

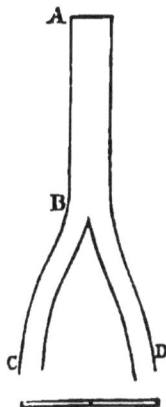

Beats. The theory of beats is given in Smith's *Harmonics*, published in 1749. Tartini's treatise, in which the sounds called by his name were first discussed, appeared in 1754.

The diatonic scale. The ordinary or diatonic scale consists of a series of notes, for which the times of vibration are in the ratio of the numbers in the following table:

$$C \quad D \quad E \quad F \quad G \quad A \quad B \quad C$$
$$1, \frac{8}{9}, \frac{4}{5}, \frac{3}{4}, \frac{2}{3}, \frac{3}{5}, \frac{8}{15}, \frac{1}{2};$$

or, in other words, the numbers of vibrations per second are in the ratio of

$$1, \frac{9}{8}, \frac{5}{4}, \frac{4}{3}, \frac{3}{2}, \frac{5}{3}, \frac{15}{8}, 2,$$

that is, of the numbers

$$24, \quad 27, \quad 30, \quad 32, \quad 36, \quad 40, \quad 45, \quad 48.$$

As a matter of fact the actual number of vibrations corresponding to the particular C employed as a central note varies in different places, and from time to time. As an ordinary standard for the concert pitch of this note C, the middle C as it is usually called, about 264 vibrations per second are taken to constitute the note, and the vibration numbers of the several notes of the diatonic scale are therefore

$$264, \quad 297, \quad 330, \quad 352, \quad 396, \quad 440, \quad 495, \quad 528.$$

Taking 1100 feet per second as the velocity of sound, the wave length for the middle C is $1100 \div 264$ which is nearly 4·17 feet. This note can therefore be produced by an open pipe 2·08 feet in length or by a stopped pipe of the length 1·04 feet.

It will be seen that if we take the ratio of each vibration number to the one preceding it, we obtain the following series

$$\frac{9}{8}, \quad \frac{10}{9}, \quad \frac{16}{15}, \quad \frac{9}{8}, \quad \frac{10}{9}, \quad \frac{9}{8}, \quad \frac{16}{15}.$$

These ratios mark the intervals between the notes, and of these intervals $\frac{9}{8}$ and $\frac{10}{9}$ are called *tones* and $\frac{16}{15}$ is called a *semi-tone*.

Chromatic scale. The diatonic scale is enlarged by the insertion of five other notes, one between the first and second, one between the second and third, one between the fourth and fifth, one between the fifth and sixth, and the other between the sixth and seventh. This scale is called the chromatic scale, and the notes inserted are such that, roughly, the interval between each consecutive pair of notes is a semi-tone.

The range of sounds appreciable by the human ear varies for different persons, but in general extends over above nine octaves. A series of aerial impulses will produce the impression of a continuous note when they recur with such rapidity that the ear cannot appreciate the succession of impulses, and it is found that this is the case for a wave length of about 68 feet. On the other hand it has been found that the highest note which is in general appreciable has about eight-fifths of an inch for its wave length.

The frequencies, or vibration numbers of these notes are about 16 and 8250.

The general case of Art. (230), *when the listener is in one train, and the note is sounded in another train, may be treated as follows*.*

Let the source of sound travel from A to B in $1/n$th of a second, n being the frequency of the note.

Let the listener be at P when the commencement of a vibration reaches him, and at Q when the vibration has completely entered his ear, and take the time from P to Q to be $1/m$th of a second.

* This mode of treatment is due to Mr A. W. Flux.

Then, u being the velocity of the source of sound and u' of the listener,

$$\frac{AP}{v} + \frac{PQ}{u'} = \text{the time from the sound leaving } A \text{ to the listener's arrival}$$

at Q

$$= \frac{AB}{u} + \frac{BQ}{v} = \frac{AB}{u} + \frac{AP + BQ - AB}{v};$$

$$\therefore PQ \left\{ \frac{1}{u'} - \frac{1}{v} \right\} = AB \left\{ \frac{1}{u} - \frac{1}{v} \right\}.$$

Hence, since $PQ = \dfrac{u'}{m}$, and $AB = \dfrac{u}{n}$,

$$\frac{m}{n} = \frac{v - u'}{v - u},$$

and therefore the frequency of the apparent note is

$$n \frac{v - u'}{v - u}.$$

This expression, as will easily be seen, includes the four cases previously given.

Algebraical representation of a vibration.

The displacement of a cycloidal pendulum from its position of rest at the time t is represented by an expression of the form

$$a \sin \left(2\pi \frac{t}{\tau} + a \right),$$

where τ is the time of a complete oscillation.

If then we are given the time of vibration, τ, for any kind of vibratory movement, and a the extreme displacement of the moving particle from its position of rest, the above expression may be employed as representing the two most important characteristics of the motion.

If n is the frequency of a note, the expression becomes

$$a \sin (2\pi nt + a).$$

In general the expression does not represent the exact position of the moving particle, except at the point of greatest velocity, and at the points of zero velocity, but it does enable us to express in a calculable form the phenomena of interferences.

The mathematical principle employed is that the resultant of any number of small displacements of the same kind is equal to their algebraical sum.

Thus the two notes $a \sin 2\pi nt$ and $a \sin (2\pi nt + \pi)$, when sounded together, destroy each other, and the result is silence, as in the experiment described on a previous page.

If the two notes $a \sin (2\pi nt + a)$ and $a \sin (2\pi nt + \beta)$ are sounded together, the result is

$$2a \cos \frac{a - \beta}{2} \sin \left(2\pi nt + \frac{a + \beta}{2} \right),$$

representing the same note with a different intensity.

In a similar manner the resultant of two notes of the same frequency, but of different intensities, will be the same note with a different intensity, such notes would be represented by the expressions

$$a \sin (2\pi nt + a) \quad \text{and} \quad b \sin (2\pi nt + \beta).$$

This however presupposes that the timbre, or quality, of each note is the same.

If the timbre is not the same, as for instance if one note is sounded on a violin, and the other on a flute, the two would be heard as separate sounds.

On the subjects of this chapter the student may consult, amongst many other books, Spencer's *Treatise on Music*, in Weale's series, Ganot's *Physics*, Deschanel's *Natural Philosophy*, Jamin's *Cours de Physique*, Tyndall's *Sound*, Sedley Taylor's *Treatise on Sound and Music*, the introductory chapter of Lord Rayleigh's *Sound*, and Helmholtz's *Tonempfindungen*.

1. A semicircle is immersed vertically in liquid with the diameter in the surface ; shew how to divide it into n sectors, such that the pressure on each is the same.

2. A sphere is totally immersed in water, and a line is drawn from the centre representing in magnitude and direction the resultant of the fluid pressure on the surface of any hemisphere ; prove that the locus of the extremity of this line is a sphere.

3. Prove that a thin uniform rod will float in a vertical position in stable equilibrium in a liquid of n times its density, if a heavy particle be attached to its lower end of weight greater than $(\sqrt{n}-1)$ times its own weight.

4. A box is made of uniform material in the form of a pyramid whose base is a regular polygon of n sides and whose slant sides are equal isosceles triangles having a common vertex; each of these triangles forms a lid which turns about a side of the polygon as a hinge, its weight is w, and its plane makes with the vertical an angle β, and when closed the box is water-tight. It is filled with water of weight W and is placed on a horizontal table, shew that nw must be greater than $\frac{3}{2} W \operatorname{cosec}^2 \beta$ or else the water will escape.

5. A plane area is completely immersed in water, its plane being vertical. It is made to descend in a vertical plane without any rotation and with uniform velocity, shew that the centre of pressure approaches the horizontal through the centre of gravity with a velocity which is inversely proportional to the square of the depth of the centre of gravity.

6. A diving bell is lowered in a lake until one half of it is filled with water ; prove that, if d is the depth of the top of the bell below the surface, the height of the bell is $2 (h-d)$, where h is the height of the water barometer.

7. If a quadrilateral area be entirely immersed in water, and if a, β, γ, δ are the depths of its four corners, and h the depth of its centroid, prove that the depth of its centre of pressure is

$$\frac{1}{2}(a+\beta+\gamma+\delta) - \frac{1}{6h}(\beta\gamma+\gamma a+a\beta+a\delta+\beta\delta+\gamma\delta).$$

8. A canal of triangular section is constructed by placing two continuous triangular prisms of concrete on a rough plane; if the sections of the prisms be equal triangles ABC, $A'B'C'$, having the angles C, C', in contact and A, A', at the surface of the water, then shew that $\tan C$ should be less than $2 \tan A$, and further the angle of friction between the concrete and the ground must be greater than $\cot^{-1}\{\rho \cot B + (1+\rho) \cot C\}$, where ρ is the density of concrete.

9. A series of conical shells made of paper of weight w per unit area have equal vertical angles $2a$, and circular rims of radii $a, 2a, \dots na$ respectively. They stand one inside another, with their rims resting on a slightly damped horizontal table. Within the smallest cone, and between successive cones, are gases of such densities that each cone is on the point of rising from the table, the damp surface of which exerts a vertical capillary force κ per unit length of rim. Find the pressures of the respective gases, and shew that if all the rims except the outside one become dry so that the gases mix, the outside cone will rise unless held down by an additional force

$$\frac{1}{12}\pi a(n-1)\{3wa\operatorname{cosec}a.n(n-1)+4\kappa(2n-1)\}.$$

10. A portion of a homogeneous elliptic cylinder, the eccentricity of a right section of which is $\frac{1}{2}$, is bounded by one of the planes through the latera recta of the cross sections and floats with its axis in the surface of a liquid and no part of the bounding plane immersed.

Prove that the density of the liquid is to that of the cylinder as $8\pi + 3\sqrt{3} : 6\pi$, and that there are three positions of equilibrium of which two are unstable.

11. If the chamber of a diving bell, of height a, could contain a weight W of water, and if the bell be lowered so that the depth of the highest point is d, prove that when the temperature absolute $T°$ is raised $t°$, the tension of the supporting chain is diminished by

$$What/aT\{(h+d)^2+4ah\}^{\frac{1}{2}}$$

nearly, h being the height of the water barometer.

12. A weightless inextensible envelope full of air floats in equilibrium in the receiver of an air-pump; find the velocity of its descent after n strokes of the piston, supposed instantaneous, and made at equal intervals.

13. If the volume of the receiver be n times that of the barrel, and if v be the limit of the above velocity when n is infinite, and v' the velocity which would have been obtained in vacuo in the same time, shew that $v' = \epsilon v$.

14. A bent tube ABC contains fluid, and the tube rotates uniformly with an angular velocity ω about the leg AB, which is vertical: find the position of equilibrium of the fluid.

If l be the whole length of tube occupied by the fluid, and the angle $ABC = a$, examine the case in which $\omega^2 > \dfrac{g}{2l} \cot^2 \dfrac{a}{2}$.

15. A spherical bubble of air ascends in water; having given its radius c at the depth $2h$, find its radius at the depth h, h being the height of the water barometer.

16. Prove that the work done in pumping air into a diving bell, over and above the work done in compressing the air varies jointly as the volume of the water displaced and the depth of the centroid of this volume below the outer surface of the water.

17. A spherical shell is partly full of water at rest. If the water be made to rotate about the vertical diameter, shew that the greatest depression of the free surface exceeds its greatest elevation.

18. Liquid of density ρ is standing in a fixed smooth circular cylinder with axis vertical and of radius a. This is made to revolve about the axis with angular velocity ω, none of the base being exposed. A paraboloidal solid of density σ shaped just to fit the cavity in the liquid is gently placed upon the surface so that its flat top just passes through the highest rim of the liquid. If $\rho > \sigma$, shew that before it reaches its equilibrium position, the liquid rising round it (supposing no interference with the base to take place), it must sink through a depth

$$\{(\rho \dot{-} \sigma)^{\frac{1}{2}} / \rho^{\frac{1}{2}} - 1\}^2 \omega^2 a^2 / 4g.$$

19. Two very small spheres, of the same size but different densities, are connected by a fine string and immersed in a liquid which rotates uniformly about a fixed axis, and is not acted upon by any external forces; find their position of relative equilibrium.

20. Close to the base of a vertical cylinder there is a small aperture turned upwards as in the figure, Art. (207), but instead of the surface in the cylinder being free, a heavy piston rests upon it; find the height to which the jet rises.

21. Find the horse-power of an engine that can, in four hours, fill a bath 50 feet in length, 20 feet in breadth, and 5 feet in depth, with water raised on an average 12 feet, and elevate its temperature 6° F.; having given $J = 772$ foot-pounds, and that the useful work is half the whole work expended by the engine.

22. Shew that, if the height of the barometer vary from one end of a lake to the other, there will result a heaping up of the water on one side. Find what will be the greatest rise above the mean level produced in a circular lake of 100 miles diameter by a variation in the height of the mercury of ·001 inch per mile.

23. The height of Niagara is 162 feet. Shew that the falling water may be made to balance a column of water 324 feet high in a J-shaped tube with its lower mouth under the fall.

24. A cylindrical bucket with water in it balances a mass M over a pulley. A piece of cork, of mass m and specific gravity σ, is then tied by a string to the bottom of the bucket so as to be wholly immersed. Prove that the tension of this string will be

$$\frac{2Mmg}{2M+m}\left(\frac{1}{\sigma}-1\right);$$

and that the pressure on the curved surface of the bucket will be greater or less than before according as the volume of the cork has to the volume of the water a ratio greater or less than

$$\sqrt{1+\frac{m}{2M}}-1:1.$$

25. Two buckets containing water, the mass of each bucket with the contained water being M, balance each other over a smooth pulley. Two pieces of wood of masses m, m', and specific gravities σ, σ' are then tied to the bottoms of the buckets so as to be wholly immersed, prove that the tension of the string attached to the mass m is

$$\frac{2m(M+m')g}{2M+m+m'}\left(\frac{1}{\sigma}-1\right).$$

26. If a jet of liquid flows out of a vessel prove that across a vena contracta of area k, at which the pressure would be increased by p if the liquid were at rest, the quantity of liquid and the amount of momentum which issue in the time t are respectively $kt\sqrt{2p\rho}$ and $2kpt$, where ρ is the density of the liquid.

27. A hollow sphere, smooth inside, is filled with liquid; if the sphere is made to revolve uniformly about a vertical axis with which it is rigidly connected, find, at any instant, the surface of equal pressure.

28. A cylindrical aperture, smooth inside, cut through a solid cylinder, with its axis parallel to that of the cylinder, is filled with liquid, and closed so that no liquid can escape; the cylinder being made to roll uniformly on a horizontal plane, find at any instant the surfaces of equal pressure, and trace their changes through a whole revolution of the cylinder.

29. If a mass of homogeneous liquid rotates about an axis and is acted upon by a force to a point in the axis varying inversely as the

square of the distance, the curvatures of the meridian curve of the free surface at the equator and pole are respectively $1/a\,(1-m)$ and $(1-mb^3/a^3)/b$, where a and b are the equatorial and polar radii and m is the ratio of the centrifugal force to the attraction at the equator.

30.   A vertical cylindrical vessel which is filled with water, has an aperture bored in its side at the depth h below the surface of the water. If K is the area of the surface of the water, and κ the area of the aperture, prove that when the surface of the water has descended through the distance x, the velocity u with which the surface is then descending will be such that

$$u^2 \{xK^2 + (h-x)\,\kappa^2\} = g\kappa^2\,\{2hx - x^2\}.$$

31.   A tuning fork held over a glass jar of a certain depth has its sound greatly augmented; but a jar an inch deeper, or an inch shallower, produces but a slight augmentation.   Why is this the case?

32.   On clapping your hands near a long railing, a sound is heard resembling that produced by the swift passage of a switch through the air; state the cause of this sound.

33.   A wooden sphere of radius r is held just immersed in a cylindrical vessel of radius R containing water, and is allowed to rise gently out of the water; prove that the loss of potential energy of the water is $Wr\,(3R^2 - 2r^2) \div 3R^2$, W being the weight of water displaced by the sphere.

If the sphere be allowed to rise until it is half out of the water, prove that the loss of potential energy is to the loss in the previous case in the ratio of

$$39R^2 - 24r^2 : 48R^2 - 32r^2.$$

If the sphere be left to itself when under water, and if we could suppose the water to come at rest on the sphere leaving it, what would be the velocity with which the sphere would shoot out?

34.   The times of the aerial vibrations constituting a note C, and its fifth G, are in the ratio $3 : 2$; compare the times of the vibrations of C and the fifth of G.

35.   A wheel with 33 teeth strikes a card in spinning, and thereby produces a note which is two octaves above the middle C; find the number of revolutions of the wheel per second.

36.   In a train, which is passing a station at the rate of 60 miles an hour, a musical note is sounded.   Assuming that the velocity of sound is 1120 feet per second find how much the pitch of the note is raised, to a listener at the station, as the train approaches the station, and how much it is lowered when the train recedes from the station.

If the note be the fifteenth of the middle C, find the vibration numbers of the apparent notes.

ANSWERS TO EXAMPLES.

CHAPTER I. Examination.

4. $10\frac{1}{2}$ lbs. wt. and 42 lbs. wt.   5. wa.

6. 180 lbs. wt.   8. 82944 lbs. wt.

CHAPTER II. Examination.

2. 1687·5, ·0361689814, 22896, ·49074.

3. 28316 and 384193 very nearly.

4. ·0352736.   5. 6·2425.   6. 6·48 cub. ft.

7. 18·52.   8. 6 and $5\frac{5}{8}$.   9. 104976.

10. 2ρ.   11. $202\frac{1}{2}$.   12. 13.

13. $3\sigma - s' - s''$.

CHAPTER II. Examples.

1. $\dfrac{mn+n}{mn+1}\rho$ and $\dfrac{m+1}{mn+1}\rho$.   2. 2ρ.

3. $\dfrac{n+1}{2}, \dfrac{2n+1}{3}, \dfrac{n+2}{3}$.   4. $3\sigma' - 2\sigma,\ 4\sigma - 3\sigma'$.

5. 3 : 1.   6. 4·762 gallons.   8. $\dfrac{1}{25}$ of a foot.

CHAPTER III. Examination.

2. $43\frac{9}{12}$ lbs. wt. per sq. inch.   About 58 lbs. wt.

4. $73\frac{13}{11}$ lbs. wt. per sq. inch, neglecting atmospheric pressure.

6. wt. of 62·5 lbs.   8. $\dfrac{2}{3} w\pi r h \sqrt{r^2 + h^2}$.

10.   If h is height of rectangle and a the depth of the upper side, the depth of the c.p. is

$$\frac{2}{3} \cdot \frac{h^2 + 3ah + 3a^2}{h + 2a}.$$

11.   Depth of line is $\dfrac{h}{\sqrt{2}}$,

12.   Depths of lines are in the ratios $1 : \sqrt{2} : \sqrt{3}\ldots\ldots$

13.   Depth of line $= \dfrac{1}{\sqrt[3]{2}} h$.

CHAPTER III. Examples.

2.   3125 lbs.

3.   The line divides the opposite side in the ratio of $3 : 1$.

4. $\dfrac{1}{8\sqrt{2}}$ (whole length of liquid).   5. $16 + \dfrac{125\pi}{4}$ lbs.

6. $20 + \dfrac{125\pi}{4}$ lbs.   8. $1 : 1$.

11.   The point lies in the line from the vertex bisecting the base and at a depth $\dfrac{1}{\sqrt{3}}$ (the depth of the vertex).

12. $1 : \sqrt{2} - 1$.   13. $\dfrac{4}{3}(1 + \sqrt{10})$ inches.

14. $1 : 4 : 9$.

17.   The increase $= 14$ (the weight of the fluid).

25.   The densities are equal.

30.   The depths are $\dfrac{1}{\sqrt[3]{2}} h$ and $\dfrac{1}{2} h$.

33. $4(5\sqrt{2} - 7) : 3$.   34. $3 : 6 : 4$.

CHAPTER IV. Examination.

7.   12 feet.   8. $16 : 15$.   9.   The forces are equal.

14. $\frac{4}{5}$ths of a cubic foot.

CHAPTER IV. Examples.

2. $\dfrac{1}{866}$ of a cubic yard.   4. $\sqrt[3]{2} - 1 : 1$.

7. $\frac{1}{4}$th of the cylinder is in the upper liquid.

8. $\frac{9}{4}$ density of wood.

9. Half that of water.

10. $\left(\frac{375\pi}{64} - 8\right)$ oz. wt.

14. $r^2 : 2(r^3 - r'^3)$.

16. One-third of the axis is immersed.

18. $h\sqrt{\dfrac{\sigma}{\rho}}$, h being the height of the paraboloid.

19. $2186\frac{2}{3}$ tons wt.

22. $\dfrac{r}{\sqrt{2}}$.

26. $19 : 56$.

27. $\dfrac{3r}{2}$.

41. $whr^2\left(1 - \dfrac{\pi}{6}\right)$.

CHAPTER V. Examination.

2. $72\cdot7$.

3. $1 : 12$.

4. $1 : 8(1+at)$ and $1 : 2(1+at)$.

7. $11\frac{3}{7}$ and $-11\frac{3}{7}$.

14. 1909 lbs. wt.

21. The weight of one-third of a ton.

24. $\cdot0004$.

CHAPTER V. Examples.

1. $1+at : n^3$.

6. $h : h'$.

8. Length above surface is changed in ratio $1 : \cdot9987$.

12. Seven times the height of the water barometer.

18. About 17 inches.

19. $1 : \sqrt{3}$.

21. 8100.

CHAPTER VI. Examination.

1. To one-third of its original volume.

4. 512 lbs. wt.

5. Early in the 4th stroke.

8. $\dfrac{5000\pi}{16}$ lbs.

9. $\dfrac{3000\pi}{16}$ lbs.

13. Three feet.

CHAPTER VI. Examples.

3. $6\cdot1$ inches nearly.

9. If a is the length originally occupied by air, h the height of the barometer, and ρ the density of atmospheric air, the difference x is given by the equation

$$x^2 + 2a(x-h) + \frac{\rho_n}{\rho}h(2a+x) = 0.$$

CHAPTER VII. EXAMINATION.

1. $5 : 7.$ 2. $9 : 7.$ 3. $1280 - \pi : 1280 - 2\pi.$

4. $\dfrac{198}{1805}$ of a cubic foot.   5. $\dfrac{25}{21}$ and $\dfrac{20}{21}.$

7. $\dfrac{4}{5}.$ 8. $\dfrac{4}{5}.$

CHAPTER VII. EXAMPLES.

2. $10 \cdot 8$ nearly.   3. $1 \cdot 9$ nearly.   4. $4 : 5.$

5. $4 : 3.$ 6. The diamond contains $5\frac{1}{4}$ grains.

7. $12\frac{1}{2}$ shillings.

8. $\gamma a' \beta' (\beta - a) + \gamma' a \beta (a' - \beta') = \gamma \gamma' (a' \beta - a \beta').$

CHAPTER VIII. EXAMINATION.

1. $15\frac{45}{1120}$ lbs. wt.   8. $75°.$

10. $\dfrac{pV + p'V'}{U} \cdot \dfrac{1 + at'}{1 + at}.$ 11. The air at greatest pressure.

CHAPTER VIII. EXAMPLES.

1. $22°\frac{58}{211}.$ 2. $13°\frac{53}{204}.$

3. The difference of the observed pressures.

CHAPTER IX.

1. Inversely as the radii.   2. $3 : 2.$

3. 80 lbs. wt. on a sq. inch.   4. $r'^3 : r^3.$

10. About $\frac{1}{15}$th inch.   11. $\frac{1}{18}$th inch.

CHAPTER X.

1. $\cdot 096$ and $\cdot 04$, $\cdot 2$, $\cdot 48.$

3. $288 : 35$, and, Π being the atmospheric pressure,

$$\left(\frac{7}{9}\right)^3 \left(\Pi + \frac{252}{35}\right) : \left(\frac{4}{3}\right)^3 \left(\Pi + \frac{7}{8}\right).$$

MISCELLANEOUS PROBLEMS IN HYDROSTATICS.

4. $4 : 3 \sin a$.

5. Distances from B of points of division are in the ratios
$$1 : \sqrt{2} : \sqrt{3} : \&c.$$

6. $\dfrac{1}{3}$ and $\dfrac{13}{36}$.   7. Densities are equal.

15. The inclination to the horizon $= \cos^{-1} 2\dfrac{1-m}{1+m}$.   23. $60°$.

24. The weight of the vessel must be at least $\dfrac{17}{7}$ (the weight of the fluid).

30. If ρ, ρ' be the densities of the lower and upper liquids respectively, σ the density of the rod, and θ its inclination to the vertical,
$$\cos^2 \theta = \frac{c^2}{a^2} \frac{\rho - \rho'}{\sigma - \rho'}.$$

42. The inclination of the radius vector to the surface is $60°$.

43. The point lies in the central generating line, dividing it in the ratio $2 : 1$.

44. In the first case the point divides the central generating line in the ratio $3 : 1$; in the second it bisects the generating line.

55. If ρ be the density of the upper and ρ' of the lower liquid, the pressures are in the ratio $4\rho : 3\rho + \rho'$.

68. $n^3 : 1 + at$.   93. $1 : 4$.

CHAPTER XIII.

1. 201250, 2990058·57, 267662·5 and, nearly, 3976778.

2. 1710·625, 820·15625.   3. 1002178 nearly.

4. 13310208π, and two-thirds of this number.

5. 834624, 1788279069.

6. 474·05, 76·2, 1014237·8496.

7. $694·232r$.   8. $10 : 3$.

CHAPTER XIV.

2. If $l =$ Latus Rectum, $\omega^2 l = g$.

3. Length submerged $= \dfrac{\sigma}{\rho} h + \dfrac{\omega^2 r^2}{4g}$.

4. A paraboloid.   6. $\dfrac{\pi a^2 \omega^2}{4g}$.

8. $\dfrac{1}{4}\pi\rho a^4\omega^2,\ \dfrac{1}{4}\pi\rho a^4\omega^2 + g\pi\rho a^2 h,\ g\pi\rho a h^2 + \pi\rho a^3 h\omega^2.$

9. $\dfrac{1}{4}\pi\rho a^4\omega^2 + \dfrac{1}{3}g\pi\rho a^3.$

CHAPTER XV.

1. The point divides the slant side in the ratio $\tan^2 a : 4.$
2. $r'h : rh'.$ 10. $2nlA/B.$ 15. $\rho v\pi r h^2.$

INDEX.

CAMBRIDGE : PRINTED BY J. & C. F. CLAY, AT THE UNIVERSITY PRESS.

A

CLASSIFIED CATALOGUE

OF

EDUCATIONAL WORKS

PUBLISHED BY

GEORGE BELL & SONS

LONDON: YORK STREET, COVENT GARDEN
NEW YORK: 66, FIFTH AVENUE; AND BOMBAY
CAMBRIDGE: DEIGHTON, BELL & CO.

DECEMBER, 1895

CONTENTS.

GREEK AND LATIN CLASSICS.

ANNOTATED AND CRITICAL EDITIONS.

AESCHYLUS. Edited by F. A. PALEY, M.A., LL.D., late Classical Exa miner to the University of London. *4th edition, revised.* 8vo, 8s.

[*Bib. Class.*

— Edited by F. A. PALEY, M.A., LL.D. 6 vols. Fcap. 8vo, 1s. 6d.

[*Camb. Texts with Notes.*

Agamemnon.	Persae.
Choephoroe.	Prometheus Vinctus.
Eumenides.	Septem contra Thebas.

ARISTOPHANIS Comoediae quae supersunt cum perditarum fragmentis tertiis curis, recognovit additis adnotatione critica, summariis, descriptione metrica, onomastico lexico HUBERTUS A. HOLDEN, LL.D. [late Fellow of Trinity College, Cambridge]. Demy 8vo.

Vol. I., containing the Text expurgated, with Summaries and Critical Notes, 18s.

The Plays sold separately :

Acharnenses, 2s.	Aves, 2s.
Equites, 1s. 6d.	Lysistrata, et Thesmophoriazu-
Nubes, 2s.	sae, 4s.
Vespae, 2s.	Ranae, 2s.
Pax, 2s.	Plutus, 2s.

Vol. II. Onomasticon Aristophaneum continens indicem geographl-cum et bistoricum. 5s. 6d.

— The Peace. A revised Text with English Notes and a Preface. By F. A. PALEY, M.A., LL.D. Post 8vo, 2s. 6d. [*Pub. Sch. Ser.*

— The Acharnians. A revised Text with English Notes and a Preface. By F. A. PALEY, M.A., LL.D. Post 8vo, 2s. 6d. [*Pub. Sch. Ser.*

— The Frogs. A revised Text with English Notes and a Preface. By F. A. PALEY, M.A., LL.D. Post 8vo, 2s. 6d. [*Pub. Sch. Ser.*

CAESAR De Bello Gallico. Edited by GEORGE LONG, M.A. *New edition.* Fcap. 8vo, 4s.

Or in parts, Books I.-III., 1s. 6d. ; Books IV. and V., 1s. 6d. ; Books VI. and VII., 1s. 6d. [*Gram. Sch. Class.*

— De Bello Gallico. Book I. Edited by GEORGE LONG, M.A. With Vocabulary by W. F. R. SHILLETO, M.A. 1s. 6d. [*Lower Form Ser.*

— De Bello Gallico. Book II. Edited by GEORGE LONG, M.A. With Vocabulary by W. F. R. SHILLETO, M.A. Fcap. 8vo, 1s. 6d.

[*Lower Form Ser.*

— De Bello Gallico. Book III. Edited by GEORGE LONG, M.A. With Vocabulary by W. F. R. SHILLETO, M.A. Fcap. 8vo, 1s. 6d.

[*Lower Form Ser.*

— Seventh Campaign in Gaul. B.C. 52. De Bello Gallico, Lib. VII. Edited with Notes, Excursus, and Table of Idioms, by REV. W. COOK-WORTHY COMPTON, M.A., Head Master of Dover College. With Illustrations from Sketches by E. T. COMPTON, Maps and Plans. *2nd edition.* Crown 8vo, 2s. 6d. net.

" A really admirable class book."—*Spectator.*

" One of the most original and interesting books which have been published in late years as aids to the study of classical literature. I think

CAESAR—*continued.*
it gives the student a new idea of the way in which a classical book may be made a living reality."—*Rev. J. E. C. Welldon*, Harrow.

— **Easy Selections from the Helvetian War.** Edited by A. M. M. STEDMAN, M.A. With Introduction, Notes and Vocabulary. 18mo, 1*s.*,
[*Primary Classics.*

CALPURNIUS SICULUS and M. AURELIUS OLYMPIUS NEMESIANUS. The Eclogues, with Introduction, Commentary, and Appendix. By C. H. KEENE, M.A. Crown 8vo, 6*s.*

CATULLUS, TIBULLUS, and PROPERTIUS. Selected Poems. Edited by the REV. A. H. WRATISLAW, late Head Master of Bury St. Edmunds School, and F. N. SUTTON, B.A. With Biographical Notices of the Poets. Fcap. 8vo, 2*s.* 6*d.* [*Gram. Sch. Class.*

CICERO'S Orations. Edited by G. LONG, M.A. 8vo. [*Bib. Class.*
Vol. I.—In Verrem. 8*s.*
Vol. II.—Pro P. Quintio—Pro Sex. Roscio—Pro Q. Roscio—Pro M. Tullio—Pro M. Fonteio—Pro A. Caecina—De Imperio Cn. Pompeii—Pro A. Cluentio—De Lege Agraria—Pro C. Rabirio. 8*s.*
Vols. III. and IV. *Out of print.*

— **De Senectute, De Amicitia, and Select Epistles.** Edited by GEORGE LONG, M.A. *New edition.* Fcap. 8vo, 3*s.* [*Gram. Sch. Class.*

— **De Amicitia.** Edited by GEORGE LONG, M.A. Fcap. 8vo, 1*s.* 6*d.*
[*Camb. Texts with Notes.*

— **De Senectute.** Edited by GEORGE LONG, M.A. Fcap. 8vo, 1*s.* 6*d.*
[*Camb. Texts with Notes.*

— **Epistolae Selectae.** Edited by GEORGE LONG, M.A. Fcap. 8vo, 1*s.* 6*d.*
[*Camb. Texts with Notes.*

— **The Letters to Atticus.** Book I. With Notes, and an Essay on the Character of the Writer. By A. PRETOR, M.A., late of Trinity College, Fellow of St. Catherine's College, Cambridge. *3rd edition.* Post 8vo, 4*s.* 6*d.* [*Pub. Sch. Ser.*

CORNELIUS NEPOS. Edited by the late REV. J. F. MACMICHAEL, Head Master of the Grammar School, Ripon. Fcap. 8vo, 2*s.*
[*Gram. Sch. Class.*

DEMOSTHENES. Edited by R. WHISTON, M.A., late Head Master of Rochester Grammar School. 2 vols. 8vo, 8*s.* each. [*Bib. Class.*
Vol. I.—Olynthiacs—Philippics—De Pace—Halonnesus—Chersonese—Letter of Philip—Duties of the State—Symmoriae—Rhodians—Megalopolitans—Treaty with Alexander—Crown.
Vol. II.—Embassy—Leptines—Meidias—Androtion—Aristocrates—Timocrates—Aristogeiton.

— **De Falsa Legatione.** By the late R. SHILLETO, M.A., Fellow of St. Peter's College, Cambridge. *8th edition.* Post 8vo, 6*s.* [*Pub. Sch. Ser.*

— **The Oration against the Law of Leptines.** With English Notes. By the late B. W. BEATSON, M.A., Fellow of Pembroke College. *3rd edition.* Post 8vo, 3*s.* 6*d.* [*Pub. Sch. Ser.*

EURIPIDES. By F. A. PALEY, M.A., LL.D. 3 vols. *2nd edition, revised.* 8vo, 8*s.* each. Vol. I. *Out of print.* [*Bib. Class.*
Vol. II.—Preface—Ion—Helena—Andromache—Electra—Bacchae—Hecuba. 2 Indexes.
Vol. III.—Preface—Hercules Furens—Phoenissae—Orestes—Iphigenia in Tauris—Iphigenia in Aulide—Cyclops. 2 Indexes.

EURIPIDES. Electra. Edited, with Introduction and Notes, by C. H. KEENE, M.A., Dublin, Ex-Scholar and Gold Medallist in Classics. Demy 8vo, 10s. 6d.
— Edited by F. A. PALEY, M.A., LL.D. 13 vols. Fcap. 8vo, 1s. 6d. each.
[*Camb. Texts with Notes.*

Alcestis.	Phoenissae.
Medea.	Troades.
Hippolytus.	Hercules Furens.
Hecuba.	Andromache.
Bacchae.	Iphigenia in Tauris.
Ion (2s.).	Supplices.
Orestes.	

HERODOTUS. Edited by REV. J. W. BLAKESLEY, B.D. 2 vols. 8vo, 12s.
[*Bib. Class.*
— Easy Selections from the Persian Wars. Edited by A. G. LIDDELL, M.A. With Introduction, Notes, and Vocabulary. 18mo, 1s. 6d.
[*Primary Classics.*

HESIOD. Edited by F. A. PALEY, M.A., LL.D. 2nd edition, revised. 8vo, 5s.
[*Bib. Class.*

HOMER. Edited by F. A. PALEY, M.A., LL.D. 2 vols. 2nd edition, revised. 14s. Vol. II. (Books XIII.-XXIV.) may be had separately. 6s.
[*Bib. Class.*
— Iliad. Books I.-XII. Edited by F. A. PALEY, M.A., LL.D. Fcap. 8vo, 4s. 6d.
Also in 2 Parts. Books I.-VI. 2s. 6d. Books VII.-XII. 2s. 6d.
[*Gram. Sch. Class.*
— Iliad. Book I. Edited by F. A. PALEY, M.A., LL.D. Fcap. 8vo, 1s.
[*Camb. Text with Notes.*

HORACE. Edited by REV. A. J. MACLEANE, M.A. 4th edition, revised by GEORGE LONG. 8vo, 8s. [*Bib. Class.*
— Edited by A. J. MACLEANE, M.A. With a short Life. Fcap. 8vo, 3s. 6d. Or, Part I., Odes, Carmen Seculare, and Epodes, 2s.; Part II., Satires, Epistles, and Art of Poetry, 2s. [*Gram. Sch. Class.*
— Odes. Book I. Edited by A. J. MACLEANE, M.A. With a Vocabulary by A. H. DENNIS, M.A. Fcap. 8vo, 1s. 6d. [*Lower Form Ser.*

JUVENAL: Sixteen Satires (expurgated). By HERMAN PRIOR, M.A., late Scholar of Trinity College, Oxford. Fcap. 8vo, 3s. 6d.
[*Gram. Sch. Class.*

LIVY. The first five Books, with English Notes. By J. PRENDEVILLE. A new edition revised throughout, and the notes in great part re-written, by J. H. FREESE, M.A., late Fellow of St. John's College, Cambridge. Books I. II. III. IV. V. With Maps and Introductions. Fcap. 8vo, 1s. 6d. each.
— Book VI. Edited by E. S. WEYMOUTH, M.A., Lond., and G. F. HAMILTON, B.A. With Historical Introduction, Life of Livy, Notes, Examination Questions, Dictionary of Proper Names, and Map. Crown 8vo, 2s. 6d.
— Book XXI. By the REV. L. D. DOWDALL, M.A., late Scholar and University Student of Trinity College, Dublin, B.D., Ch. Ch. Oxon. Post 8vo, 2s. [*Pub. Sch. Ser.*
— Book XXII. Edited by the REV. L. D. DOWDALL, M.A., B.D. Post 8vo, 2s. [*Pub. Sch. Ser.*

LIVY. Easy Selections from the Kings of Rome. Edited by A. M. M. STEDMAN, M.A. With Introduction, Notes, and Vocabulary. 18mo, 1s. 6d. [*Primary Class.*

LUCAN. The Pharsalia. By C. E. HASKINS, M.A., Fellow of St. John's College, Cambridge, with an Introduction by W. E. HEITLAND, M.A., Fellow and Tutor of St. John's College, Cambridge. 8vo, 14s.

LUCRETIUS. Titi Lucreti Cari De Rerum Natura Libri Sex. By the late H. A. J. MUNRO, M.A., Fellow of Trinity College, Cambridge. *4th edition, finally revised.* 3 vols. Demy 8vo. Vols. I., II., Introduction, Text, and Notes, 18s. Vol. III., Translation, 6s.

MARTIAL: Select Epigrams. Edited by F. A. PALEY, M.A., LL.D., and the late W. H. STONE, Scholar of Trinity College, Cambridge. With a Life of the Poet. Fcap. 8vo, 4s. 6d. [*Gram. Sch. Class.*

OVID: Fasti. Edited by F. A. PALEY, M.A., LL.D. *Second edition.* Fcap. 8vo, 3s. 6d. [*Gram. Sch. Class.*
Or in 3 vols, 1s. 6d. each [*Grammar School Classics*], or 2s. each [*Camb. Texts with Notes*], Books I. and II., Books III. and IV., Books V. and VI.

— Selections from the Amores, Tristia, Heroides, and Metamorphoses. By A. J. MACLEANE, M.A. Fcap. 8vo, 1s. 6d. [*Camb. Texts with Notes.*

Ars Amatoria et Amores. A School Edition. Carefully Revised and Edited, with some Literary Notes, by J. HERBERT WILLIAMS, M.A., late Demy of Magdalen College, Oxford. Fcap. 8vo, 3s. 6d.

Heroides XIV. Edited, with Introductory Preface and English Notes, by ARTHUR PALMER, M.A., Professor of Latin at Trinity College, Dublin. Demy 8vo, 6s.

— Metamorphoses, Book XIII. A School Edition. With Introduction and Notes, by CHARLES HAINES KEENE, M.A., Dublin, Ex-Scholar and Gold Medallist in Classics. *3rd edition.* Fcap. 8vo, 2s. 6d.

— Epistolarum ex Ponto Liber Primus. With Introduction and Notes, by CHARLES HAINES KEENE, M.A. Crown 8vo, 3s.

PLATO. The Apology of Socrates and Crito. With Notes, critical and exegetical, by WILHELM WAGNER, PH.D. *12th edition.* Post 8vo, 3s. 6d. A CHEAP EDITION. Limp Cloth. 2s. 6d. [*Pub. Sch. Ser.*

— Phaedo. With Notes, critical and exegetical, and an Analysis, by WILHELM WAGNER, PH.D. *10th edition.* Post 8vo, 5s. 6d. [*Pub. Sch. Ser.*

— Protagoras. The Greek Text revised, with an Analysis and English Notes, by W. WAYTE, M.A., Classical Examiner at University College, London. *7th edition.* Post 8vo, 4s. 6d. [*Pub. Sch. Ser.*

— Euthyphro. With Notes and Introduction by G. H. WELLS, M.A., Scholar of St. John's College, Oxford; Assistant Master at Merchant Taylors' School. *3rd edition.* Post 8vo, 3s. [*Pub. Sch. Ser.*

— The Republic. Books I. and II. With Notes and Introduction by G. H. WELLS, M.A. *4th edition,* with the Introduction re-written. Post 8vo, 5s. [*Pub. Sch. Ser.*

— Euthydemus. With Notes and Introduction by G. H. WELLS, M.A. Post 8vo, 4s. [*Pub. Sch. Ser.*

— Phaedrus. By the late W. H. THOMPSON, D.D., Master of Trinity College, Cambridge. 8vo, 5s. [*Bib. Class.*

— Gorgias. By the late W. H. THOMPSON, D.D., Master of Trinity College, Cambridge. *New edition.* 6s. [*Pub. Sch. Ser.*

PLAUTUS. Aulularia. With Notes, critical and exegetical, by W. WAGNER, PH.D. *5th edition.* Post 8vo, 4*s.* 6*d.* [*Pub. Sch. Ser.*

— **Trinummus.** With Notes, critical and exegetical, by WILHELM WAGNER, PH.D. *5th edition.* Post 8vo, 4*s.* 6*d.* [*Pub. Sch. Ser.*

— **Menaechmei.** With Notes, critical and exegetical, by WILHELM WAGNER, PH.D. *2nd edition.* Post 8vo, 4*s.* 6*d.* [*Pub. Sch. Ser.*

— **Mostellaria.** By E. A. SONNENSCHEIN, M.A., Professor of Classics at Mason College, Birmingham. Post 8vo, 5*s.* [*Pub. Sch. Ser.*

— **Captivi.** Abridged and Edited for the Use of Schools. With Introduction and Notes by J. H. FREESE, M.A., formerly Fellow of St. John's College, Cambridge. Fcap. 8vo, 1*s.* 6*d.*

PROPERTIUS. Sex. Aurelii Propertii Carmina. The Elegies of Propertius, with English Notes. By F. A. PALEY, M.A., LL.D. *2nd edition.* 8vo, 5*s.*

SALLUST: Catilina and Jugurtha. Edited, with Notes, by the late GEORGE LONG. *New edition, revised,* with the addition of the Chief Fragments of the Histories, by J. G. FRAZER, M.A., Fellow of Trinity College, Cambridge. Fcap. 8vo, 3*s.* 6*d.*, or separately, 2*s.* each.
[*Gram. Sch. Class.*

SOPHOCLES. Edited by REV. F. H. BLAYDES, M.A. Vol. I. Oedipus Tyrannus—Oedipus Coloneus—Antigone. 8vo, 8*s.* [*Bib. Class.*
Vol. II. Philoctetes—Electra—Trachiniae—Ajax. By F. A. PALEY, M.A., LL.D. 8vo, 6*s.*, or the four Plays separately in limp cloth, 2*s.* 6*d.* each.

— **Trachiniae.** With Notes and Prolegomena. By ALFRED PRETOR, M.A., Fellow of St. Catherine's College, Cambridge. Post 8vo, 4*s.* 6*d.*
[*Pub. Sch. Ser.*

— **The Oedipus Tyrannus of Sophocles.** By B. H. KENNEDY, D.D., Regius Professor of Greek and Hon. Fellow of St. John's College, Cambridge. With a Commentary containing a large number of Notes selected from the MS. of the late T. H. STEEL, M.A. Crown 8vo, 8*s.*

— — A SCHOOL EDITION. Post 8vo, 2*s.* 6*d.* [*Pub. Sch. Ser.*

— Edited by F. A. PALEY, M.A., LL.D. 5 vols. Fcap. 8vo, 1*s.* 6*d.* each.
[*Camb. Texts with Notes.*

Oedipus Tyrannus.	Electra.
Oedipus Coloneus.	Ajax.
Antigone.	

TACITUS: Germania and Agricola. Edited by the late REV. P. FROST, late Fellow of St. John's College, Cambridge. Fcap. 8vo, 2*s.* 6*d.*
[*Gram. Sch. Class.*

— **The Germania.** Edited, with Introduction and Notes, by R. F. DAVIS, M.A. Fcap. 8vo, 1*s.* 6*d.*

TERENCE. With Notes, critical and explanatory, by WILHELM WAGNER, PH.D. *3rd edition.* Post 8vo, 7*s.* 6*d.* [*Pub. Sch. Ser.*

— Edited by WILHELM WAGNER, PH.D. 4 vols. Fcap. 8vo, 1*s.* 6*d.* each.
[*Camb. Texts with Notes.*

Andria.	Hautontimorumenos.
Adelphi.	Phormio.

THEOCRITUS. With short, critical and explanatory Latin Notes, by F. A. PALEY, M.A., LL.D. *2nd edition, revised.* Post 8vo, 4*s.* 6*d.*
[*Pub. Sch. Ser.*

THUCYDIDES, Book VI. By T. W. DOUGAN, M.A., Fellow of St. John's College, Cambridge; Professor of Latin in Queen's College, Belfast. Edited with English notes. Post 8vo, 2s. [*Pub. Sch. Ser.*

— The History of the Peloponnesian War. With Notes and a careful Collation of the two Cambridge Manuscripts, and of the Aldine and Juntine Editions. By the late RICHARD SHILLETO, M.A., Fellow of St. Peter's College, Cambridge. 8vo. Book I. 6s. 6d. Book II. 5s. 6d.

VIRGIL. By the late PROFESSOR CONINGTON, M.A. Revised by the late PROFESSOR NETTLESHIP, Corpus Professor of Latin at Oxford. 8vo.
[*Bib. Class.*

Vol. I. The Bucolics and Georgics, with new Memoir and three Essays on Virgil's Commentators, Text, and Critics. *4th edition.* 10s. 6d.
Vol. II. The Aeneid, Books I.-VI. *4th edition.* 10s. 6d.
Vol. III. The Aeneid, Books VII.-XII. *3rd edition.* 10s. 6d.

— Abridged from PROFESSOR CONINGTON'S Edition, by the REV. J. G. SHEPPARD, D.C.L., H. NETTLESHIP, late Corpus Professor of Latin at the University of Oxford, and W. WAGNER, PH.D. 2 vols. Fcap. 8vo, 4s. 6d. each. [*Gram. Sch. Class.*
Vol. I. Bucolics, Georgics, and Aeneid, Books I.-IV.
Vol. II. Aeneid, Books V.-XII.
Also the Bucolics and Georgics, in one vol. 3s.

Or in 9 separate volumes (*Grammar School Classics, with Notes at foot of page*), *price* 1s. 6d. *each.*

Bucolics.	Aeneid, V. and VI.
Georgics, I. and II.	Aeneid, VII. and VIII.
Georgics, III. and IV.	Aeneid, IX. and X.
Aeneid, I. and II.	Aeneid, XI. and XII.
Aeneid, III. and IV.	

Or in 12 *separate volumes* (*Cambridge Texts with Notes at end*), *price* 1s. 6d. *each.*

Bucolics.	Aeneid, VII.
Georgics, I. and II.	Aeneid, VIII.
Georgics, III. and IV.	Aeneid, IX.
Aeneid, I. and II.	Aeneid, X.
Aeneid, III. and IV.	Aeneid, XI.
Aeneid, V. and VI. (price 2s.)	Aeneid, XII.

Aeneid, Book I. CONINGTON'S Edition abridged. With Vocabulary by W. F. R. SHILLETO, M.A. Fcap. 8vo, 1s. 6d. [*Lower Form Ser.*

XENOPHON : Anabasis. With Life, Itinerary, Index, and three Maps. Edited by the late J. F. MACMICHAEL. *Revised edition.* Fcap. 8vo, 3s. 6d. [*Gram. Sch. Class.*
Or in 4 *separate volumes, price* 1s. 6d. *each.*
Book I. (with Life, Introduction, Itinerary, and three Maps)—Books II. and III.—Books IV. and V.—Books VI. and VII.

— Anabasis. MACMICHAEL'S Edition, revised by J. E. MELHUISH, M.A., Assistant Master of St. Paul's School. In 6 volumes, fcap. 8vo. With Life, Itinerary, and Map to each volume, 1s. 6d. each.
[*Camb. Texts with Notes.*
Book I.—Books II. and III.—Book IV.—Book V.—Book VI.—Book VII.

XENOPHON. Cyropaedia. Edited by G. M. GORHAM, M.A., late Fellow of Trinity College, Cambridge. *New edition.* Fcap. 8vo, 3*s*. 6*d*.
[*Gram. Sch. Class.*
    Also Books I. and II., 1*s*. 6*d*. ; Books V. and VI., 1*s*. 6*d*.
— Memorabilia. Edited by PERCIVAL FROST, M.A., late Fellow of St. John's College, Cambridge. Fcap. 8vo, 3*s*. [*Gram. Sch. Class.*
— Hellenica. Book I. Edited by L. D. DOWDALL, M.A., B.D. Fcap. 8vo, 2*s*. [*Camb. Texts with Notes.*
— Hellenica. Book II. By L. D. DOWDALL, M.A., B.D. Fcap. 8vo, 2*s*.
[*Camb. Texts with Notes.*

TEXTS.

AESCHYLUS. Ex novissima recensione F. A. PALEY, A.M., LL.D. Fcap. 8vo, 2*s*. [*Camb. Texts.*
CAESAR De Bello Gallico. Recognovit G. LONG, A.M. Fcap. 8vo, 1*s*. 6*d*. [*Camb. Texts.*
CATULLUS. A New Text, with Critical Notes and an Introduction, by J. P. POSTGATE, M.A., LITT.D., Fellow of Trinity College, Cambridge, Professor of Comparative Philology at the University of London. Wide fcap. 8vo, 3*s*.
CICERO De Senectute et de Amicitia, et Epistolae Selectae. Recensuit G. LONG, A.M. Fcap. 8vo, 1*s*. 6*d*. [*Camb. Texts.*
CICERONIS Orationes in Verrem. Ex recensione G. LONG, A.M. Fcap. 8vo, 2*s*. 6*d*. [*Camb. Texts.*
CORPUS POETARUM LATINORUM, a se aliisque denuo recognitorum et brevi lectionum varietate instructorum, edidit JOHANNES PERCIVAL POSTGATE. Tom. I.—Ennius, Lucretius, Catullus, Horatius, Vergilius, Tibullus, Propertius, Ovidius. Large post 4to, 21*s*. net. Also in 2 Parts, sewed, 9*s*. each, net.
  .*.* To be completed in 4 parts, making 2 volumes.
CORPUS POETARUM LATINORUM. Edited by WALKER. Containing :—Catullus, Lucretius, Virgilius, Tibullus, Propertius, Ovidius, Horatius, Phaedrus, Lucanus, Persius, Juvenalis, Martialis, Sulpicia, Statius, Silius Italicus, Valerius Flaccus, Calpurnius Siculus, Ausonius, and Claudianus. 1 vol. 8vo, cloth, 18*s*.
EURIPIDES. Ex recensione F. A. PALEY, A.M., LL.D. 3 vols. Fcap. 8vo, 2*s*. each. [*Camb. Texts.*
  Vol. I.—Rhesus—Medea—Hippolytus—Alcestis—Heraclidae—Supplices—Troades.
  Vol. II.—Ion—Helena—Andromache—Electra—Bacchae—Hecuba.
  Vol. III.—Hercules Furens—Phoenissae—Orestes—Iphigenia in Tauris —Iphigenia in Aulide—Cyclops.
HERODOTUS. Recensuit J. G. BLAKESLEY, S.T.B. 2 vols. Fcap. 8vo, 2*s*. 6*d*. each. [*Camb. Texts.*
HOMERI ILIAS I.-XII. Ex novissima recensione F. A. PALEY, A.M., LL.D. Fcap. 8vo, 1*s*. 6*d*. [*Camb. Texts.*
HORATIUS. Ex recensione A. J. MACLEANE, A.M. Fcap. 8vo, 1*s*. 6*d*.
[*Camb. Texts.*
JUVENAL ET PERSIUS. Ex recensione A. J. MACLEANE, A.M. Fcap. 8vo, 1*s*. 6*d*. [*Camb. Texts.*

LUCRETIUS. Recognovit H. A. J. MUNRO, A.M. Fcap. 8vo, 2s.
[Camb. Texts.

PROPERTIUS. Sex. Propertii Elegiarum Libri IV. recensuit A. PALMER, collegii sacrosanctae et individuae Trinitatis juxta Dublinum Socius. Fcap. 8vo, 3s. 6d.
— Sexti Properti Carmina. Recognovit JOH. PERCIVAL POSTGATE, Large post 4to, boards, 3s. 6d. net.

SALLUSTI CRISPI CATILINA ET JUGURTHA, Recognovit G. LONG, A.M. Fcap. 8vo, 1s. 6d. *[Camb. Texts.*

SOPHOCLES. Ex recensione F. A. PALEY, A.M., LL.D. Fcap. 8vo, 2s. 6d.
[Camb. Texts.

TERENTI COMOEDIAE. GUL. WAGNER relegit et emendavit. Fcap. 8vo, 2s. *[Camb. Texts.*

THUCYDIDES. Recensuit J. G. DONALDSON, S.T.P. 2 vols. Fcap. 8vo, 2s. each. *[Camb. Texts.*

VERGILIUS. Ex recensione J. CONINGTON, A.M. Fcap. 8vo, 2s.
[Camb. Texts.

XENOPHONTIS EXPEDITIO CYRI. Recensuit J. F. MACMICHAEL. A.B. Fcap. 8vo, 1s. 6d. *[Camb. Texts.*

TRANSLATIONS.

AESCHYLUS, The Tragedies of. Translated into English verse by ANNA SWANWICK. *4th edition revised.* Small post 8vo, 5s.
— The Tragedies of. Literally translated into Prose, by T. A. BUCKLEY, B.A. Small post 8vo, 3s. 6d.
— The Tragedies of. Translated by WALTER HEADLAM, M.A., Fellow of King's College, Cambridge. · *[Preparing.*

ANTONINUS (M. Aurelius), The Thoughts of. Translated by GEORGE LONG, M.A. *Revised edition.* Small post 8vo, 3s. 6d.
Fine paper edition on handmade paper. Pott 8vo, 6s.

APOLLONIUS RHODIUS. The Argonautica. Translated by E. P. COLERIDGE. Small post 8vo, 5s.

AMMIANUS MARCELLINUS. History of Rome during the Reigns of Constantius, Julian, Jovianus, Valentinian, and Valens. Translated by PROF. C. D. YONGE, M.A. With a complete Index. Small post 8vo, 7s. 6d.

ARISTOPHANES, The Comedies of. Literally translated by W. J. HICKIE. *With Portrait.* 2 vols. Small post 8vo, 5s. each.
Vol. I.—Acharnians, Knights, Clouds, Wasps, Peace, and Birds.
Vol. II.—Lysistrata, Thesmophoriazusae, Frogs, Ecclesiazusae, and Plutus.
— The Acharnians. Translated by W. H. COVINGTON, B.A. With Memoir and Introduction. Crown 8vo, sewed, 1s.

ARISTOTLE on the Athenian Constitution. Translated, with Notes and Introduction, by F. G. KENYON, M.A., Fellow of Magdalen College, Oxford. Pott 8vo, printed on handmade paper. *2nd edition.* 4s. 6d.
— History of Animals. Translated by RICHARD CRESSWELL, M.A. Small post 8vo, 5s.

ARISTOTLE. Organon: or, Logical Treatises, and the Introduction of Porphyry. With Notes, Analysis, Introduction, and Index, by the REV. O. F. OWEN, M.A. 2 vols. Small post 8vo, 3s. 6d. each.
— Rhetoric and Poetics. Literally Translated, with Hobbes' Analysis, &c., by T. BUCKLEY, B.A. Small post 8vo, 5s.
— Nicomachean Ethics. Literally Translated, with Notes, an Analytical Introduction, &c., by the Venerable ARCHDEACON BROWNE, late Classical Professor of King's College. Small post 8vo, 5s.
— Politics and Economics. Translated, with Notes, Analyses, and Index, by E. WALFORD, M.A., and an Introductory Essay and a Life by DR. GILLIES. Small post 8vo, 5s.
— Metaphysics. Literally Translated, with Notes, Analysis, &c., by the REV. JOHN H. M'MAHON, M.A. Small post 8vo, 5s.
ARRIAN. Anabasis of Alexander, together with the Indica. Translated by E. J. CHINNOCK, M.A., LL.D. With Introduction, Notes, Maps, and Plans. Small post 8vo, 5s.
CAESAR. Commentaries on the Gallic and Civil Wars, with the Supplementary Books attributed to Hirtius, including the complete Alexandrian, African, and Spanish Wars. Translated by W. A. M'DEVITTE, B.A. Small post 8vo, 5s.
— Gallic War. Translated by W. A. M'DEVITTE, B.A. 2 vols., with Memoir and Map. Crown 8vo, sewed. Books I. to IV., Books V. to VII., 1s. each.
CALPURNIUS SICULUS, The Eclogues of. The Latin Text, with English Translation by E. J. L. SCOTT, M.A. Crown 8vo, 3s. 6d.
CATULLUS, TIBULLUS, and the Vigil of Venus. Prose Translation. Small post 8vo, 5s.
CICERO, The Orations of. Translated by PROF. C. D. YONGE, M.A. With Index. 4 vols. Small post 8vo, 5s. each.
— On Oratory and Orators. With Letters to Quintus and Brutus. Translated by the REV. J. S. WATSON, M.A. Small post 8vo, 5s.
— On the Nature of the Gods. Divination, Fate, Laws, a Republic, Consulship. Translated by PROF. C. D. YONGE, M.A., and FRANCIS BARHAM. Small post 8vo, 5s.
— Academics, De Finibus, and Tusculan Questions. By PROF. C. D. YONGE, M.A. Small post 8vo, 5s.
— Offices; or, Moral Duties. Cato Major, an Essay on Old Age; Laelius, an Essay on Friendship; Scipio's Dream; Paradoxes; Letter to Quintus on Magistrates. Translated by C. R. EDMONDS. *With Portrait,* 3s. 6d.
— Old Age and Friendship. Translated, with Memoir and Notes, by G. H. WELLS, M.A. Crown 8vo, sewed, 1s.
DEMOSTHENES, The Orations of. Translated, with Notes, Arguments, a Chronological Abstract, Appendices, and Index, by C. RANN KENNEDY. 5 vols. Small post 8vo.
    Vol. I.—The Olynthiacs, Philippics. 3s. 6d.
    Vol. II.—On the Crown and on the Embassy. 5s.
    Vol. III.—Against Leptines, Midias, Androtion, and Aristocrates. 5s.
    Vols. IV. and V.—Private and Miscellaneous Orations. 5s. each.
— On the Crown. Translated by C. RANN KENNEDY. Crown 8vo, sewed, 1s.
DIOGENES LAERTIUS. Translated by PROF. C. D. YONGE, M.A. Small post 8vo, 5s.

EPICTETUS, The Discourses of. With the Encheiridion and Fragments. Translated by GEORGE LONG, M.A. Small post 8vo, 5s.
Fine Paper Edition, 2 vols. Pott 8vo, 10s. 6d.

EURIPIDES. A Prose Translation, from the Text of Paley. By E. P. COLERIDGE, B.A. 2 vols., 5s. each.
Vol. I.—Rhesus, Medea, Hippolytus, Alcestis, Heraclidæ, Supplices, Troades, Ion, Helena.
Vol. II.—Andromache, Electra, Bacchæ, Hecuba, Hercules Furens, Phoenissae, Orestes, Iphigenia in Tauris, Iphigenia in Aulis, Cyclops.
.*. The plays separately (except Rhesus, Helena, Electra, Iphigenia in Aulis, and Cyclops). Crown 8vo, sewed, 1s. each.
— Translated from the Text of Dindorf. By T. A. BUCKLEY, B.A. 2 vols. small post 8vo, 5s. each.

GREEK ANTHOLOGY. Translated by GEORGE BURGES, M.A. Small post 8vo, 5s.

HERODOTUS. Translated by the REV. HENRY CARY, M.A. Small post 8vo, 3s. 6d.
— Analysis and Summary of. By J. T. WHEELER. Small post 8vo, 5s.

HESIOD, CALLIMACHUS, and THEOGNIS. Translated by the REV. J. BANKS, M.A. Small post 8vo, 5s.

HOMER. The Iliad. Translated by T. A. BUCKLEY, B.A. Small post 8vo, 5s.
— The Odyssey, Hymns, Epigrams, and Battle of the Frogs and Mice. Translated by T. A. BUCKLEY, B.A. Small post 8vo, 5s.
— The Iliad. Books I.-IV. Translated into English Hexameter Verse, by HENRY SMITH WRIGHT, B.A., late Scholar of Trinity College, Cambridge. Medium 8vo, 5s.

HORACE. Translated by Smart. *Revised edition.* By T. A. BUCKLEY, B.A. Small post 8vo, 3s. 6d.
— The Odes and Carmen Saeculare. Translated into English Verse by the late JOHN CONINGTON, M.A., Corpus Professor of Latin in the University of Oxford. 11th edition. Fcap. 8vo. 3s. 6d.
— The Satires and Epistles. Translated into English Verse by PROF. JOHN CONINGTON, M.A. 8th edition. Fcap. 8vo, 3s. 6d.
— Odes and Epodes. Translated by SIR STEPHEN E. DE VERE, BART. 3rd edition, enlarged. Imperial 16mo. 7s. 6d. net.

ISOCRATES, The Orations of. Translated by J. H. FREESE, M.A., late Fellow of St. John's College, Cambridge, with Introductions and Notes. Vol. I. Small post 8vo, 5s.

JUSTIN, CORNELIUS NEPOS, and EUTROPIUS. Translated by the REV. J. S. WATSON, M.A. Small post 8vo, 5s.

JUVENAL, PERSIUS, SULPICIA, and LUCILIUS. Translated by L. EVANS, M.A. Small post 8vo, 5s.

LIVY. The History of Rome. Translated by DR. SPILLAN, C. EDMONDS, and others. 4 vols. small post 8vo, 5s. each.
— Books I., II., III., IV. A Revised Translation by J. H. FREESE, M.A., late Fellow of St. John's College, Cambridge. With Memoir, and Maps. 4 vols., crown 8vo, sewed, 1s. each.
— Book V. and Book VI. A Revised Translation by E. S. WEYMOUTH, M.A., Lond. With Memoir, and Maps. Crown 8vo, sewed, 1s. each.
— Book IX. Translated by FRANCIS STORR, B.A. With Memoir. Crown 8vo, sewed, 1s.

LUCAN. The Pharsalia. Translated into Prose by H. T. RILEY. Small post 8vo, 5s.
— **The Pharsalia.** Book I. Translated by FREDERICK CONWAY, M.A. With Memoir and Introduction. Crown 8vo, sewed, 1s.

LUCIAN'S Dialogues of the Gods, of the Sea-Gods, and of the Dead. Translated by HOWARD WILLIAMS, M.A. Small post 8vo, 5s.

LUCRETIUS. Translated by the REV. J. S. WATSON, M.A. Small post 8vo, 5s.
— Literally translated by the late H. A. J. MUNRO, M.A. *4th edition.* Demy 8vo, 6s.

MARTIAL'S Epigrams, complete. Literally translated into Prose, with the addition of Verse Translations selected from the Works of English Poets, and other sources. Small post 8vo, 7s. 6d.

OVID, The Works of. Translated. 3 vols. Small post 8vo, 5s. each.
Vol. I.—Fasti, Tristia, Pontic Epistles, Ibis, and Halieuticon.
Vol. II.—Metamorphoses. *With Frontispiece.*
Vol. III.—Heroides, Amours, Art of Love, Remedy of Love, and Minor Pieces. *With Frontispiece.*
— **Fasti.** Translated by H. T. RILEY, B.A. 3 vols. Crown 8vo, sewed, 1s. each.
— **Tristia.** Translated by H. T. RILEY, B.A. Crown 8vo, sewed, 1s.

PINDAR. Translated by DAWSON W. TURNER. Small post 8vo, 5s.

PLATO. Gorgias. Translated by the late E. M. COPE, M.A., Fellow of Trinity College. *2nd edition.* 8vo, 7s.
— **Philebus.** Translated by F. A. PALEY, M.A., LL.D. Small 8vo, 4s.
— **Theaetetus.** Translated by F. A. PALEY, M.A., LL.D. Small 8vo, 4s.
— **The Works of.** Translated, with Introduction and Notes. 6 vols. Small post 8vo, 5s. each.
Vol. I.—The Apology of Socrates—Crito—Phaedo—Gorgias—Protagoras—Phaedrus—Theaetetus—Eutyphron—Lysis. Translated by the REV. H. CARY.
Vol. II.—The Republic—Timaeus—Critias. Translated by HENRY DAVIS.
Vol. III.—Meno—Euthydemus—The Sophist—Statesman—Cratylus—Parmenides—The Banquet. Translated by G. BURGES.
Vol. IV.—Philebus—Charmides—Laches—Menexenus—Hippias—Ion—The Two Alcibiades—Theages—Rivals—Hipparchus—Minos—Clitopho—Epistles. Translated by G. BURGES.
Vol. V.—The Laws. Translated by G. BURGES.
Vol. VI.—The Doubtful Works. Edited by G. BURGES. With General Index to the six volumes.
— **Apology, Crito, Phaedo, and Protagoras.** Translated by the REV. H. CARY. Small post 8vo, sewed, 1s., cloth, 1s. 6d.
— **Dialogues.** A Summary and Analysis of. With Analytical Index, giving references to the Greek text of modern editions and to the above translations. By A. DAY, LL.D. Small post 8vo, 5s.

PLAUTUS, The Comedies of. Translated by H. T. RILEY, B.A. 2 vols. Small post 8vo, 5s. each.
Vol. I.—Trinummus—Miles Gloriosus—Bacchides—Stichus—Pseudolus—Menaechmei—Aulularia—Captivi—Asinaria—Curculio.
Vol. II.—Amphitryon—Rudens—Mercator—Cistellaria—Truculentus—Persa—Casina—Poenulus—Epidicus—Mostellaria—Fragments.
— **Trinummus, Menaechmei, Aulularia, and Captivi.** Translated by H. T. RILEY, B.A. Small post 8vo, sewed, 1s., cloth, 1s. 6d.

PLINY. The Letters of Pliny the Younger. Melmoth's Translation, revised, by the REV. F. C. T. BOSANQUET, M.A. Small post 8vo, 5s.

PLUTARCH. Lives. Translated by A. STEWART, M.A., late Fellow of Trinity College, Cambridge, and GEORGE LONG, M.A. 4 vols. small post 8vo, 3s. 6d. each.

— Morals. Theosophical Essays. Translated by C. W. KING, M.A., late Fellow of Trinity College, Cambridge. Small post 8vo, 5s.

— Morals. Ethical Essays. Translated by the REV. A. R. SHILLETO, M.A. Small post 8vo, 5s.

PROPERTIUS. Translated by REV. P. J. F. GANTILLON, M.A., and accompanied by Poetical Versions, from various sources. Small post 8vo, 3s. 6d.

PRUDENTIUS, Translations from. A Selection from his Works, with a Translation into English Verse, and an Introduction and Notes, by FRANCIS ST. JOHN THACKERAY, M.A., F.S.A., Vicar of Mapledurham, formerly Fellow of Lincoln College, Oxford, and Assistant-Master at Eton. Wide post 8vo, 7s. 6d.

QUINTILIAN: Institutes of Oratory, or, Education of an Orator. Translated by the REV. J. S. WATSON, M.A. 2 vols. small post 8vo, 5s. each.

SALLUST, FLORUS, and **VELLEIUS PATERCULUS.** Translated by J. S. WATSON, M.A. Small post 8vo, 5s.

SENECA: On Benefits. Translated by A. STEWART, M.A., late Fellow of Trinity College, Cambridge. Small post 8vo, 3s. 6d.

— Minor Essays and On Clemency. Translated by A. STEWART, M.A. Small post 8vo, 5s.

SOPHOCLES. Translated, with Memoir, Notes, etc., by E. P. COLERIDGE, B.A. Small post 8vo, 5s.
Or the plays separately, crown 8vo, sewed, 1s. each.

— The Tragedies of. The Oxford Translation, with Notes, Arguments, and Introduction. Small post 8vo, 5s.

— The Dramas of. Rendered in English Verse, Dramatic and Lyric, by SIR GEORGE YOUNG, BART., M.A., formerly Fellow of Trinity College, Cambridge. 8vo, 12s. 6d.

— The Œdipus Tyrannus. Translated into English Prose. By PROF. B. H. KENNEDY. Crown 8vo, in paper wrapper, 1s.

SUETONIUS. Lives of the Twelve Caesars and Lives of the Grammarians. Thomson's revised Translation, by T. FORESTER. Small post 8vo, 5s.

TACITUS, The Works of. Translated, with Notes and Index 2 vols.. Small post 8vo, 5s. each.
Vol. I.—The Annals.
Vol. II.—The History, Germania, Agricola, Oratory, and Index.

TERENCE and **PHAEDRUS.** Translated by H. T. RILEY, B.A. Small post 8vo, 5s.

THEOCRITUS, BION, MOSCHUS, and **TYRTAEUS.** Translated by the REV. J. BANKS, M.A. Small post 8vo, 5s.

THEOCRITUS. Translated into English Verse by C. S. CALVERLEY, M.A., late Fellow of Christ's College, Cambridge. *New edition, revised.* Crown 8vo, 7s. 6d.

THUCYDIDES. The Peloponnesian War. Translated by the REV. H.
DALE. *With Portrait.* 2 vols., 3*s.* 6*d.* each.
— Analysis and Summary of. By J. T. WHEELER. Small post 8vo, 5*s.*
VIRGIL. Translated by A. HAMILTON BRYCE, LL.D. With Memoir and
Introduction. Small post 8vo, 3*s.* 6*d.*
Also in 6 vols., crown 8vo, sewed, 1*s.* each.

Georgics.	Æneid IV.-VI.
Bucolics.	Æneid VII.-IX.
Æneid I.-III.	· Æneid X.-XII.

XENOPHON. The Works of. In 3 vols. Small post 8vo, 5*s.* each.
Vol. I.—The Anabasis, and Memorabilia. Translated by the REV. J. S.
WATSON, M.A. With a Geographical Commentary, by W. F. AINSWORTH,
F.S.A. F.R.G.S., etc.
Vol. II.—Cyropaedia and Hellenics. Translated by the REV. J. S.
WATSON, M.A., and the REV. H. DALE.
Vol. III.—The Minor Works. Translated by the REV. J. S.
WATSON, M.A.
— Anabasis. Translated by the REV. J. S. WATSON, M.A. With Memoir
and Map. 3 vols.
— Hellenics. Books I. and II. Translated by the REV. H. DALE, M.A.
With Memoir.

SABRINAE COROLLA In Hortulis Regiae Scholae Salopiensis con-
texuerunt tres viri floribus legendis. *4th edition, revised and re-arranged.*
By the late BENJAMIN HALL KENNEDY, D.D., Regius Professor of Greek
at the University of Cambridge. Large post 8vo, 10*s.* 6*d.*
SERTUM CARTHUSIANUM Floribus trium Seculorum Contextum.
Cura GULIELMI HAIG BROWN, Scholae Carthusianae Archididascali.
Demy 8vo, 5*s.*
TRANSLATIONS into English and Latin. By C. S. CALVERLEY, M.A.,
late Fellow of Christ's College, Cambridge. *3rd edition.* Crown 8vo,
7*s.* 6*d.*
TRANSLATIONS from and into the Latin, Greek and English. By
R. C. JEBB, M.A., Regius Professor of Greek in the University of Cam-
bridge, H. JACKSON, M.A., LITT. D., Fellows of Trinity College, Cam-
bridge, and W. E. CURREY, M.A., formerly Fellow of Trinity College,
Cambridge. Crown 8vo. *2nd edition, revised.* 8*s.*

GRAMMAR AND COMPOSITION.

BADDELEY. Auxilia Latina. A Series of Progressive Latin Exercises.
By M. J. B. BADDELEY, M.A. Fcap. 8vo. Part I., Accidence. *5th*
edition. 2*s.* Part II. *5th edition.* 2*s.* Key to Part II. 2*s.* 6*d.*
BAIRD. Greek Verbs. A Catalogue of Verbs, Irregular and Defective;
their leading formations, tenses in use, and dialectic inflexions, with a
copious Appendix, containing Paradigms for conjugation, Rules for
formation of tenses, &c., &c. By J. S. BAIRD, T.C.D. *New edition, re-*
vised. 2*s.* 6*d.*
— Homeric Dialect. Its Leading Forms and Peculiarities. By J. S. BAIRD,
T.C.D. *New edition, revised.* By the REV. W. GUNION RUTHERFORD,
M.A., LL.D., Head Master at Westminster School. 1*s.*

BAKER. Latin Prose for London Students. By ARTHUR BAKER, M.A., Classical Master, Independent College, Taunton. Fcap. 8vo, 2s.

BARRY. Notes on Greek Accents. By the RIGHT REV. A. BARRY, D.D. *New edition, re-written.* 1s.

CHURCH. Latin Prose Lessons. By A. J. CHURCH, M.A., Professor of Latin at University College, London. *9th edition.* Fcap. 8vo, 2s. 6d.

CLAPIN. Latin Primer. By the REV. A. C. CLAPIN, M.A., Assistant Master at Sherborne School. *4th edition.* Fcap. 8vo, 1s.

COLLINS. Latin Exercises and Grammar Papers. By T. COLLINS, M.A., Head Master of the Latin School, Newport, Salop. *7th edition.* Fcap. 8vo, 2s. 6d.

— Unseen Papers in Latin Prose and Verse. With Examination Questions. *7th edition.* Fcap. 8vo, 2s. 6d.

— Unseen Papers in Greek Prose and Verse. With Examination Questions. *4th edition.* Fcap. 8vo, 3s.

— Easy Translations from Nepos, Caesar, Cicero, Livy, &c., for Retranslation into Latin. With Notes. 2s.

COMPTON. Rudiments of Attic Construction and Idiom. An Introduction to Greek Syntax for Beginners who have acquired some knowledge of Latin. By the REV. W. COOKWORTHY COMPTON, M.A., Head Master of Dover College. Crown 8vo, 3s.

FROST. Eclogae Latinae; or, First Latin Reading Book. With Notes and Vocabulary by the late REV. P. FROST, M.A. Fcap. 8vo, 1s. 6d.

— Analecta Graeca Minora. With Notes and Dictionary. *New edition.* Fcap. 8vo, 2s.

— Materials for Latin Prose Composition. By the late REV. P. FROST, M.A. *New edition.* Fcap. 8vo. 2s. Key. 4s. net.

— A Latin Verse Book. *New edition.* Fcap. 8vo, 2s. Key. 5s. net.

— Materials for Greek Prose Composition. *New edition.* Fcap. 8vo, 2s. 6d. Key. 5s. net.

— Greek Accidence. *New edition.* 1s.

— Latin Accidence. 1s.

HARKNESS. A Latin Grammar. By ALBERT HARKNESS. Post 8vo, 6s.

KEY. A Latin Grammar. By the late T. H. KEY, M.A., F.R.S. *6th thousand.* Post 8vo, 8s.

— A Short Latin Grammar for Schools. *16th edition.* Post 8vo, 3s. 6d.

HOLDEN. Foliorum Silvula. Part I. Passages for Translation into Latin Elegiac and Heroic Verse. By H. A. HOLDEN, LL.D. *11th edition.* Post 8vo, 7s. 6d.

— Foliorum Silvula. Part II. Select Passages for Translation into Latin Lyric and Comic Iambic Verse. *3rd edition.* Post 8vo, 5s.

— Foliorum Centuriae. Select Passages for Translation into Latin and Greek Prose. *10th edition.* Post 8vo, 8s.

JEBB, JACKSON, and CURREY. Extracts for Translation in Greek, Latin, and English. By R. C. JEBB, LITT.D., LL.D., Regius Professor of Greek in the University of Cambridge; H. JACKSON, LITT.D., Fellow of Trinity College, Cambridge; and W. E. CURREY, M.A., late Fellow of Trinity College, Cambridge. 4s. 6d.

Latin Syntax, Principles of. 1s.

Latin Versification. 1s.

MASON. Analytical Latin Exercises By C. P. MASON, B.A. *4th edition.* Part I., 1*s.* 6*d.* Part II., 2*s.* 6*d.*

— The Analysis of Sentences Applied to Latin. Post 8vo, 1*s.* 6*d.*

NETTLESHIP. Passages for Translation into Latin Prose. Preceded by Essays on :—I. Political and Social Ideas. II. Range of Metaphorical Expression. III. Historical Development of Latin Prose Style in Antiquity. IV. Cautions as to Orthography. By H. NETTLESHIP, M.A., late Corpus Professor of Latin in the University of Oxford. Crown 8vo, 3*s.* A Key, 4*s.* 6*d.* net.

Notabilia Quaedam; or the Principal Tenses of most of the Irregular Greek Verbs, and Elementary Greek, Latin, and French Constructions. *New edition.* 1*s.*

PALEY. Greek Particles and their Combinations according to Attic Usage. A Short Treatise. By F. A. PALEY, M.A., LL.D. 2*s.* 6*d.*

PENROSE. Latin Elegiac Verse, Easy Exercises in. By the REV. J. PENROSE. *New edition.* 2*s.* (Key, 3*s.* 6*d.* net.)

PRESTON. Greek Verse Composition. By G. PRESTON, M.A. *5th edition.* Crown 8vo, 4*s.* 6*d.*

PRUEN. Latin Examination Papers. Comprising Lower, Middle, and Upper School Papers, and a number of the Woolwich and Sandhurst Standards. By G. G. PRUEN, M.A., Senior Classical Master in the Modern Department, Cheltenham College. Crown 8vo, 2*s.* 6*d.*

SEAGER. Faciliora. An Elementary Latin Book on a New Principle. By the REV. J. L. SEAGER, M.A. 2*s.* 6*d.*

STEDMAN (A. M. M.). First Latin Lessons. By A. M. M. STEDMAN, M.A., Wadham College, Oxford. *2nd edition, enlarged.* Crown 8vo, 2*s.*

— Initia Latina. Easy Lessons on Elementary Accidence. *2nd edition.* Fcap. 8vo, 1*s.*

— First Latin Reader. With Notes adapted to the Shorter Latin Primer and Vocabulary. Crown 8vo, 1*s.* 6*d.*

— Easy Latin Passages for Unseen Translation. *2nd and enlarged edition.* Fcap. 8vo, 1*s.* 6*d.*

— Exempla Latina. First Exercises in Latin Accidence. With Vocabulary. Crown 8vo, 1*s.* 6*d.*

— The Latin Compound Sentence; Rules and Exercises. Crown 8vo, 1*s.* 6*d.* With Vocabulary, 2*s.*

— Easy Latin Exercises on the Syntax of the Shorter and Revised Latin Primers. With Vocabulary. *3rd edition.* Crown 8vo, 2*s.* 6*d.*

— Latin Examination Papers in Miscellaneous Grammar and Idioms. *3rd edition.* 2*s.* 6*d.* Key (for Tutors only), 6*s.* net.

— Notanda Quaedam. Miscellaneous Latin Exercises. On Common Rules and Idioms. *2nd edition.* Fcap. 8vo 1*s.* 6*d.* With Vocabulary, 2*s.*

— Latin Vocabularies for Repetition. Arranged according to Subjects. *3rd edition.* Fcap. 8vo, 1*s.* 6*d.*

— Steps to Greek. 18mo, 1*s.* 6*d.*

— Easy Greek Passages for Unseen Translation. Fcap. 8vo, 1*s.* 6*d.*

— Easy Greek Exercises on Elementary Syntax. [*In preparation.*

— Greek Vocabularies for Repetition. Fcap. 8vo, 1*s.* 6*d.*

— Greek Testament Selections for the Use of Schools. *2nd edition.* With Introduction, Notes, and Vocabulary. Fcap. 8vo, 2*s.* 6*d.*

— Greek Examination Papers in Miscellaneous Grammar and Idioms. *2nd edition.* 2*s.* 6*d.* Key (for Tutors only), 6*s.* net.

THACKERAY. Anthologia Graeca. A Selection of Greek Poetry, with Notes. By F. ST. JOHN THACKERAY. *5th edition.* 16mo, 4s. 6d.

— Anthologia Latina. A Selection of Latin Poetry, from Naevius tc Boëthius, with Notes. By REV. F. ST. JOHN THACKERAY. *6th edition.* 16mo, 4s. 6d.

— Hints and Cautions on Attic Greek Prose Composition. Crown 8vo, 3s. 6d.

— Exercises on the Irregular and Defective Greek Verbs. 1s. 6d.

WELLS. Tales for Latin Prose Composition. With Notes and Vocabulary. By G. H. WELLS, M.A., Assistant Master at Merchant Taylor's School. Fcap. 8vo, 2s.

HISTORY, GEOGRAPHY, AND REFERENCE BOOKS, ETC.

TEUFFEL'S History of Roman Literature. *5th edition*, revised by DR. SCHWABE, translated by PROFESSOR G. C. W. WARR, M.A, King's College, London Medium 8vo. 2 vols. 30s. Vol. I. (The Republican Period), 15s. Vol. II. (The Imperial Period), 15s.

KEIGHTLEY'S Mythology of Ancient Greece and Italy. *4th edition*, revised by the late LEONHARD SCHMITZ, PH.D., LL.D., Classical Examiner to the University of London With 12 Plates. Small post 8vo, 5s.

DONALDSON'S Theatre of the Greeks. *10th edition.* Small post 8vo, 5s.

DICTIONARY OF LATIN AND GREEK QUOTATIONS; including Proverbs, Maxims, Mottoes, Law Terms and Phrases. With all the Quantities marked, and English Translations. With Index Verborum. Small post 8vo, 5s.

A GUIDE TO THE CHOICE OF CLASSICAL BOOKS. By J. B. MAYOR, M.A., Professor of Moral Philosophy at King's College, late Fellow and Tutor of St. John's College, Cambridge. *3rd edition*, with Supplementary List. Crown 8vo, 4s. 6d.

PAUSANIAS' Description of Greece. Newly translated, with Notes and Index, by A. R. SHILLETO, M.A. 2 vols. Small post 8vo, 5s. each.

STRABO'S Geography. Translated by W. FALCONER, M.A., and H. C. HAMILTON. 3 vols. Small post 8vo, 5s. each.

AN ATLAS OF CLASSICAL GEOGRAPHY. By W. HUGHES and G. LONG, M.A. Containing Ten selected Maps. Imp. 8vo, 3s.

AN ATLAS OF CLASSICAL GEOGRAPHY. Twenty-four Maps by W. HUGHES and GEORGE LONG, M.A. With coloured outlines. Imperial 8vo, 6s.

ATLAS OF CLASSICAL GEOGRAPHY. 22 large Coloured Maps. With a complete Index. Imp. 8vo, chiefly engraved by the Messrs. Walker. 7s. 6d.

MATHEMATICS.

ARITHMETIC AND ALGEBRA.

BARRACLOUGH (T.). The Eclipse Mental Arithmetic. By TITUS BARRACLOUGH, Board School, Halifax. Standards I., II., and III., sewed, 6*d.* ; Standards II., III., and IV., sewed, 6*d.* net ; Book III., Part A, sewed, 4*d.* ; Book III., Part B, cloth, 1*s.* 6*d.*

BEARD (W. S.). Graduated Exercises in Addition (Simple and Compound). For Candidates for Commercial Certificates and Civil Service appointments. By W. S. BEARD, F.R.G.S., Head Master of the Modern School, Fareham. 3*rd edition.* Fcap. 4to, 1*s.*
— *See* PENDLEBURY.

ELSEE (C.). Arithmetic. By the REV. C. ELSEE, M.A., late Fellow of St. John's College, Cambridge, Senior Mathematical Master at Rugby School. 14*th edition.* Fcap. 8vo, 3*s.* 6*d.*
[*Camb. School and College Texts.*
— Algebra. By the REV. C. ELSEE, M.A. 8*th edition.* Fcap. 8vo, 4*s.*
[*Camb. S. and C. Texts.*

FILIPOWSKI (H. E.). Anti-Logarithms, A Table of. By H. E. FILIPOWSKI. 3*rd edition.* 8vo, 15*s.*

GOUDIE (W. P.). *See* Watson.

HATHORNTHWAITE (J. T.). Elementary Algebra for Indian Schools. By J. T. HATHORNTHWAITE, M.A., Principal and Professor of Mathematics at Elphinstone College, Bombay. Crown 8vo, 2*s.*

MACMICHAEL (W. F.) and PROWDE SMITH (R.). Algebra. A Progressive Course of Examples. By the REV. W. F. MACMICHAEL, and R. PROWDE SMITH, M.A. 4*th edition.* Fcap. 8vo, 3*s.* 6*d.* With answers, 4*s.* 6*d.*
[*Camb. S. and C. Texts.*

MATHEWS (G. B.). Theory of Numbers. An account of the Theories of Congruencies and of Arithmetical Forms. By G. B. MATHEWS, M.A., Professor of Mathematics in the University College of North Wales. Part I. Demy 8vo, 12*s.*

MOORE (B. T). Elementary Treatise on Mensuration. By B. T. MOORE, M.A., Fellow of Pembroke College, Cambridge. *New edition.* 3*s.* 6*d.*

PENDLEBURY (C.). Arithmetic. With Examination Papers and 8,000 Examples. By CHARLES PENDLEBURY, M.A., F.R.A.S., Senior Mathematical Master of St. Paul's, Author of "Lenses and Systems of Lenses, treated after the manner of Gauss." 8*th edition.* Crown 8vo. Complete, with or without Answers, 4*s.* 6*d.* In Two Parts, with or without Answers, 2*s.* 6*d.* each.
Key to Part II. 7*s.* 6*d.* net. [*Camb. Math. Ser.*
— Examples in Arithmetic. Extracted from Pendlebury's Arithmetic. With or without Answers. 6*th edition.* Crown 8vo, 3*s.*, or in Two Parts, 1*s.* 6*d.* and 2*s.* [*Camb. Math. Ser.*
— Examination Papers in Arithmetic. Consisting of 140 papers, each containing 7 questions ; and a collection of 357 more difficult problems 3*rd edition.* Crown 8vo, 2*s.* 6*d.* Key, for Tutors only, 5*s.* net.

PENDLEBURY (C.) and TAIT (T. S.). Arithmetic for Indian Schools. By C. PENDLEBURY, M.A. and T. S. TAIT, M.A., B.SC., Principal of Baroda College. Crown 8vo, 3*s*.   [*Camb. Math. Ser.*

PENDLEBURY (C.) and BEARD (W. S.). Arithmetic for the Standards. By C. PENDLEBURY, M.A., F.R.A.S., and W. S. BEARD, F.R.G.S. Standards I., II., III., sewed, 2*d.* each, cloth, 3*d.* each : IV., V., VI., sewed, 3*d.* each, cloth, 4*d.* each ; VII., sewed, 6*d.*, cloth, 8*d.* Answers to I. and II., 4*d.*, III.-VII., 4*d.* each.

— Elementary Arithmetic. *3rd edition.* Crown 8vo, 1*s.* 6*d.*

POPE (L. J.). Lessons in Elementary Algebra. By L. J. POPE, B.A. (Lond.), Assistant Master at the Oratory School, Birmingham. First Series, up to and including Simple Equations and Problems. Crown 8vo, 1*s* 6*d.*

PROWDE SMITH (R.). *See* Macmichael.

SHAW (S. J. D.). Arithmetic Papers. Set in the Cambridge Higher Local Examination, from June, 1869, to June, 1887, inclusive, reprinted by permission of the Syndicate. By S. J. D. SHAW, Mathematical Lecturer of Newnham College. Crown 8vo, 2*s.* 6*d.* ; Key, 4*s.* 6*d.* net.

TAIT (T. S.). *See* Pendlebury.

WATSON (J.) and GOUDIE (W. P.). Arithmetic. A Progressive Course of Examples. With Answers. By J. WATSON, M.A., Corpus Christi College, Cambridge, formerly Senior Mathematical Master of the Ordnance School, Carshalton. *7th edition, revised and enlarged.* By W. P. GOUDIE, B.A. Lond. Fcap. 8vo, 2*s.* 6*d.*   [*Camb. S. and C. Texts.*

WHITWORTH (W. A.). Algebra. Choice and Chance. An Elementary Treatise on Permutations, Combinations, and Probability, with 640 Exercises and Answers. By W. A. WHITWORTH, M.A., Fellow of St. John's College, Cambridge. *4th edition, revised and enlarged.* Crown 8vo, 6*s.*   [*Camb. Math. Ser.*

WRIGLEY (A.) Arithmetic. By A. WRIGLEY, M.A., St. John's College. Fcap. 8vo, 3*s.* 6*d.*   [*Camb. S. and C. Texts.*

BOOK-KEEPING.

CRELLIN (P.). A New Manual of Book-keeping, combining the Theory and Practice, with Specimens of a set of Books. By PHILLIP CRELLIN, Chartered Accountant. Crown 8vo, 3*s.* 6*d.*

— Book-keeping for Teachers and Pupils. Crown 8vo, 1*s.* 6*d.* Key, 2*s.* net.

FOSTER (B. W.). Double Entry Elucidated. By B. W. FOSTER. *14th edition.* Fcap. 4to, 3*s.* 6*d.*

MEDHURST (J. T.). Examination Papers in Book-keeping. Compiled by JOHN T. MEDHURST, A.K.C., F.S.S., Fellow of the Society of Accountants and Auditors, and Lecturer at the City of London College. *3rd edition.* Crown 8vo, 3*s.*

THOMSON (A. W.). A Text-Book of the Principles and Practice of Book-keeping. By PROFESSOR A. W. THOMSON, B.SC., Royal Agricultural College, Cirencester. *2nd edition, revised.* Crown 8vo, 5*s.*

GEOMETRY AND EUCLID.

BESANT (W. H.). Conic Sections treated Geometrically. By W. H. BESANT, SC.D., F.R.S., Fellow of St. John's College, Cambridge. *9th edition.* Crown 8vo, 4*s.* 6*d.* net. Key, 5*s.* net. [*Camb. Math. Ser.*

BRASSE (J.). The Enunciations and Figures of Euclid, prepared for Students in Geometry. By the REV. J. BRASSE, D.D. *New edition.* Fcap. 8vo, 1*s.* Without the Figures, 6*d.*

DEIGHTON (H.). Euclid. Books I.-VI., and part of Book XI., newly translated from the Greek Text, with Supplementary Propositions, Chapters on Modern Geometry, and numerous Exercises. By HORACE DEIGHTON, M.A., Head Master of Harrison College, Barbados. *3rd edition.* 4*s.* 6*d.*, or Books I.-IV., 3*s.* Books V.-XI., 2*s.* 6*d.* Key, 5*s.* net. [*Camb. Math. Ser.*

Also issued in parts :—Book I., 1*s.* ; Books I. and II., 1*s.* 6*d.* ; Books I.-III., 2*s.* 6*d.* ; Books III. and IV., 1*s.* 6*d.*

DIXON (E. T.). The Foundations of Geometry. By EDWARD T. DIXON, late Royal Artillery. Demy 8vo, 6*s.*

MASON (C. P.). Euclid. The First Two Books Explained to Beginners. By C. P. MASON, B.A. *2nd edition.* Fcap. 8vo, 2*s.* 6*d.*

McDOWELL (J.) Exercises on Euclid and in Modern Geometry, containing Applications of the Principles and Processes of Modern Pure Geometry. By the late J. McDOWELL, M.A., F.R.A.S., Pembroke College, Cambridge, and Trinity College, Dublin. *4th edition.* 6*s.* [*Camb. Math. Ser.*

TAYLOR (C.). An Introduction to the Ancient and Modern Geometry of Conics, with Historical Notes and Prolegomena. 15*s.*

— The Elementary Geometry of Conics. By C. TAYLOR, D.D., Master of St. John's College. *7th edition, revised.* With a Chapter on the Line Infinity, and a new treatment of the Hyperbola. Crown 8vo, 4*s.* 6*d.* [*Camb. Math. Ser.*

WEBB (R.). The Definitions of Euclid. With Explanations and Exercises, and an Appendix of Exercises on the First Book by R. WEBB, M.A. Crown 8vo, 1*s.* 6*d.*

WILLIS (H. G.). Geometrical Conic Sections. An Elementary Treatise. By H. G. WILLIS, M.A., Clare College, Cambridge, Assistant Master of Manchester Grammar School. Crown 8vo, 5*s.* [*Camb. Math. Ser.*

ANALYTICAL GEOMETRY, ETC.

ALDIS (W. S.). Solid Geometry, An Elementary Treatise on. By W. S. ALDIS, M.A., late Professor of Mathematics in the University College, Auckland, New Zealand. *4th edition, revised.* Crown 8vo, 6*s.* [*Camb. Math. Ser.*

BESANT (W. H.). Notes on Roulettes and Glissettes. By W. H. BESANT, SC.D., F.R.S. *2nd edition, enlarged.* Crown 8vo, 5*s.* [*Camb. Math. Ser.*

CAYLEY (A.). Elliptic Functions, An Elementary Treatise on. By
ARTHUR CAYLEY, Sadlerian Professor of Pure Mathematics in the University of Cambridge. *2nd edition.* Demy 8vo. 15*s.*

TURNBULL (W. P.). Analytical Plane Geometry, An Introduction
to. By W. P. TURNBULL, M.A., sometime Fellow of Trinity College.
8vo, 12*s.*

VYVYAN (T. G.). Analytical Geometry for Schools. By REV. T.
VYVYAN, M.A., Fellow of Gonville and Caius College, and Mathematical
Master of Charterhouse. *6th edition.* 8vo, 4*s.* 6*d.* [*Camb. S. and C. Texts.*

— Analytical Geometry for Beginners. Part I. The Straight Line and
Circle. Crown 8vo, 2*s.* 6*d.* [*Camb. Math. Ser.*

WHITWORTH (W. A.). Trilinear Co-ordinates, and other methods
of Modern Analytical Geometry of Two Dimensions. By W. A. WHIT-
WORTH, M.A., late Professor of Mathematics in Queen's College, Liverpool, and Scholar of St. John's College, Cambridge. 8vo, 16*s.*

TRIGONOMETRY.

DYER (J. M.) and WHITCOMBE (R. H.). Elementary Trigonometry. By J. M. DYER, M.A. (Senior Mathematical Scholar at Oxford),
and REV. R. H. WHITCOMBE, Assistant Masters at Eton College. *2nd
edition.* Crown 8vo, 4*s.* 6*d.* [*Camb. Math. Ser.*

PENDLEBURY (C.). Elementary Trigonometry. By CHARLES
PENDLEBURY, M.A., F.R.A.S., Senior Mathematical Master at St. Paul's
School. Crown 8vo, 4*s.* 6*d.* [*Camb. Math. Ser.*

VYVYAN (T. G.). Introduction to Plane Trigonometry. By the
REV. T. G. VYVYAN, M.A., formerly Fellow of Gonville and Caius College,
Senior Mathematical Master of Charterhouse. *3rd edition, revised and
augmented.* Crown 8vo, 3*s.* 6*d.* [*Camb. Math. Ser.*

WARD (G. H.). Examination Papers in Trigonometry. By G. H.
WARD, M.A., Assistant Master at St. Paul's School. Crown 8vo, 2*s.* 6*d.*
Key, 5*s.* net.

MECHANICS AND NATURAL PHILOSOPHY.

ALDIS (W. S.). Geometrical Optics, An Elementary Treatise on. By
W. S. ALDIS, M.A. *4th edition.* Crown 8vo, 4*s.* [*Camb. Math. Ser.*

— An Introductory Treatise on Rigid Dynamics. Crown 8vo, 4*s.*
[*Camb. Math. Ser.*

— Fresnel's Theory of Double Refraction, A Chapter on. *2nd edition,
revised.* 8vo, 2*s.*

BASSET (A. B.). A Treatise on Hydrodynamics, with numerous
Examples. By A. B. BASSET, M.A., F.R.S., Trinity College, Cambridge.
Demy 8vo. Vol. I., price 10*s.* 6*d.* ; Vol. II., 12*s.* 6*d.*

— An Elementary Treatise on Hydrodynamics and Sound. Demy
8vo, 7*s.* 6*d.*

— A Treatise on Physical Optics. Demy 8vo, 16*s.*

BESANT (W. H.). Elementary Hydrostatics. By W. H. BESANT,
SC.D., F.R.S. *16th edition.* Crown 8vo, 4*s.* 6*d.* Solutions, 5*s.* net.
[*Camb. Math. Ser.*

— Hydromechanics, A Treatise on. Part I. Hydrostatics. *5th edition
revised, and enlarged.* Crown 8vo, 5*s.* [*Camb. Math. Ser.*

BESANT (W. H.). A Treatise on Dynamics. *2nd edition.* Crown 8vo, 10s. 6d. [*Camb. Math. Ser.*

CHALLIS (PROF.). Pure and Applied Calculation. By the late REV. J. CHALLIS, M.A., F.R.S., &c. Demy 8vo, 15s.

— Physics, The Mathematical Principle of. Demy 8vo, 5s.

— Lectures on Practical Astronomy. Demy 8vo, 10s.

EVANS (J. H.) and MAIN (P. T.). Newton's Principia, The First Three Sections of, with an Appendix; and the Ninth and Eleventh Sections. By J. H. EVANS, M.A., St. John's College. The *5th edition*, edited by P. T. MAIN, M.A., Lecturer and Fellow of St. John's College. Fcap. 8vo, 4s. [*Camb. S. and C. Texts.*

GALLATLY (W.). Elementary Physics, Examples and Examination Papers in. Statics, Dynamics, Hydrostatics, Heat, Light, Chemistry, Electricity, London Matriculation, Cambridge B.A., Edinburgh, Glasgow, South Kensington, Cambridge Junior and Senior Papers, and Answers. By W. GALLATLY, M.A., Pembroke College, Cambridge, Assistant Examiner, London University. Crown 8vo, 4s. [*Camb. Math. Ser.*

GARNETT (W.). Elementary Dynamics for the use of Colleges and Schools. By WILLIAM GARNETT, M.A., D.C.L., Fellow of St. John's College, late Principal of the Durham College of Science, Newcastle-upon-Tyne. *5th edition, revised.* Crown 8vo, 6s. [*Camb. Math. Ser.*

— Heat, An Elementary Treatise on. *6th edition, revised.* Crown 8vo, 4s. 6d. [*Camb. Math. Ser.*

GOODWIN (H.). Statics. By H. GOODWIN, D.D., late Bishop of Carlisle. *2nd edition.* Fcap. 8vo, 3s. [*Camb. S. and C. Texts.*

HOROBIN (J. C.). Elementary Mechanics. Stage I. II. and III., 1s. 6d. each. By J. C. HOROBIN, M.A., Principal of Homerton New College, Cambridge.

— Theoretical Mechanics. Division I. Crown 8vo, 2s. 6d.

  . This book covers the ground of the Elementary Stage of Division I. of Subject VI. of the "Science Directory," and is intended for the examination of the Science and Art Department.

JESSOP (C. M.). The Elements of Applied Mathematics. Including Kinetics, Statics and Hydrostatics. By C. M. JESSOP, M.A., late Fellow of Clare College, Cambridge, Lecturer in Mathematics in the Durham College of Science, Newcastle-on-Tyne. Crown 8vo, 6s.
[*Camb. Math. Ser.*

MAIN (P. T.). Plane Astronomy, An Introduction to. By P. T. MAIN, M.A., Lecturer and Fellow of St. John's College. *6th edition, revised.* Fcap. 8vo, 4s. [*Camb. S. and C. Texts.*

PARKINSON (R. M.). Structural Mechanics. By R. M. PARKINSON, ASSOC. M.I.C.E. Crown 8vo, 4s. 6d.

PENDLEBURY (C.). Lenses and Systems of Lenses, Treated after the Manner of Gauss. By CHARLES PENDLEBURY, M.A., F.R.A.S., Senior Mathematical Master of St. Paul's School, late Scholar of St. John's College, Cambridge. Demy 8vo, 5s.

STEELE (R. E.). Natural Science Examination Papers. By R. E. STEELE, M.A., F.C.S., Chief Natural Science Master, Bradford Grammar School. Crown 8vo. Part I., Inorganic Chemistry, 2s. 6d. Part II., Physics (Sound, Light, Heat, Magnetism, Electricity), 2s. 6d.
[*School Exam. Series.*

WALTON (W.). **Theoretical Mechanics, Problems in.** By W. WALTON, M.A, Fellow and Assistant Tutor of Trinity Hall, Mathematical Lecturer at Magdalene College. *3rd edition, revised.* Demy 8vo, 16s.

— **Elementary Mechanics, Problems in.** *2nd edition.* Crown 8vo, 6s.
[*Camb. Math. Ser.*

DAVIS (J. F.). **Army Mathematical Papers.** Being Ten Years' Woolwich and Sandhurst Preliminary Papers. Edited, with Answers, by J. F. DAVIS, D.LIT., M.A. Lond. Crown 8vo, 2s. 6d.

DYER (J. M.) and PROWDE SMITH (R.). **Mathematical Examples.** A Collection of Examples in Arithmetic, Algebra, Trigonometry, Mensuration, Theory of Equations, Analytical Geometry, Statics, Dynamics, with Answers, &c. For Army and Indian Civil Service Candidates. By J. M. DYER, M.A., Assistant Master, Eton College (Senior Mathematical Scholar at Oxford), and R. PROWDE SMITH, M.A. Crown 8vo, 6s. [*Camb. Math. Ser.*

GOODWIN (H.). **Problems and Examples,** adapted to "Goodwin's Elementary Course of Mathematics." By T. G. VYVYAN, M.A. *3rd edition.* 8vo, 5s. ; Solutions, *3rd edition,* 8vo, 9s.

SMALLEY (G. R.). **A Compendium of Facts and Formulae in Pure Mathematics and Natural Philosophy.** By G. R. SMALLEY, F.R.A.S. *New edition, revised and enlarged.* By J. McDOWELL, M.A., F.R.A.S. Fcap. 8vo, 2s.

WRIGLEY (A.). **Collection of Examples and Problems in Arithmetic, Algebra, Geometry, Logarithms, Trigonometry, Conic Sections, Mechanics,** &c., with Answers and Occasional Hints. By the REV. A. WRIGLEY. *10th edition, 20th thousand.* Demy 8vo, 3s. 6d.

A Key. By J. C. PLATTS, M.A. and the REV. A. WRIGLEY. *2nd edition.* Demy 8vo, 5s. net.

MODERN LANGUAGES.

ENGLISH.

ADAMS (E.). **The Elements of the English Language.** By ERNEST ADAMS, PH.D. *26th edition.* Revised by J. F. DAVIS, D.LIT., M.A., (LOND.). Post 8vo, 4s. 6d.

— **The Rudiments of English Grammar and Analysis.** By ERNEST ADAMS, PH.D. *19th thousand.* Fcap. 8vo, 1s.

ALFORD (DEAN). **The Queen's English:** A Manual of Idiom and Usage. By the late HENRY ALFORD, D.D., Dean of Canterbury. *6th edition.* Small post 8vo. Sewed, 1s., cloth, 1s. 6d.

ASCHAM'S Scholemaster. Edited by PROFESSOR J. E. B. MAYOR. Small post 8vo, sewed, 1s.

BELL'S ENGLISH CLASSICS. A New Series, Edited for use in Schools, with Introduction and Notes. Crown 8vo.

> BACON'S **Essays Modernized.** Edited by F. J. ROWE, M.A., Professor of English Literature at Presidency College, Calcutta. [*Preparing.*
> BROWNING'S **Strafford.** Edited by E. H. HICKEY. With Introduction by
> S R. GARDINER, LL.D. 2s. 6d.

BELL'S ENGLISH CLASSICS—*continued.*

BURKE'S Letters on a Regicide Peace. I. and II. Edited by H. G. KEENE, M.A., C.I.E. 3*s.* ; sewed, 2*s.*

BYRON'S Childe Harold. Edited by H. G. KEENE, M.A., C.I.E., Author of "A Manual of French Literature," etc. 3*s.* 6*d.* Also Cantos I. and II. separately ; sewed, 1*s.* 9*d.*

— Siege of Corinth. Edited by P. HORDERN, late Director of Public Instruction in Burma. 1*s.* 6*d.* ; sewed, 1*s.*

CHAUCER, SELECTIONS FROM. Edited by J. B. BILDERBECK, B.A., Professor of English Literature, Presidency College, Madras. 2*s.* 6*d.* ; sewed, 1*s.* 9*d.*

DE QUINCEY'S Revolt of the Tartars and The English Mail-Coach. Edited by CECIL M. BARROW, M.A., Principal of Victoria College, Palghât, and MARK HUNTER, B.A., Principal of Coimbatore College. 3*s.* ; sewed, 1*s.*

DE QUINCEY'S Opium Eater. Edited by MARK HUNTER, B.A. [*In the press.*

GOLDSMITH'S Good-Natured Man and She Stoops to Conquer. Edited by K. DEIGHTON. Each, 2*s.* cloth ; 1*s.* 6*d.* sewed. The two plays together, sewed, 2*s.* 6*d.*

IRVING'S Sketch Book. Edited by R. G. OXENHAM, M.A. Sewed, 1*s.* 6*d.*

JOHNSON'S Life of Addison. Edited by F. RYLAND, Author of "The Students Handbook of Psychology," etc. 2*s.* 6*d.*

— Life of Swift. Edited by F. RYLAND, M.A. 2*s.*

— Life of Pope. Edited by F. RYLAND, M.A. 2*s.* 6*d.*

— Life of Milton. Edited by F. RYLAND, M.A. 2*s.* 6*d.*

— Life of Dryden. Edited by F. RYLAND, M.A. 2*s.* 6*d.*

LAMB'S Essays. Selected and Edited by K. DEIGHTON. 3*s.* ; sewed, 2*s.*

LONGFELLOW'S Evangeline. Edited by M. T. QUINN, M.A. [*In the press.*

MACAULAY'S Lays of Ancient Rome. Edited by P. HORDERN. 2*s.* 6*d.* ; sewed, 1*s.* 9*d.*

— Essay on Clive. Edited by CECIL BARROW, M.A. 2*s.* ; sewed, 1*s.* 6*d.*

MASSINGER'S A New Way to Pay Old Debts. Edited by K. DEIGHTON. 3*s.* ; sewed, 2*s.*

MILTON'S Paradise Lost. Books III. and IV. Edited by R. G. OXENHAM, M.A., Principal of Elphinstone College, Bombay. 2*s.* ; sewed, 1*s.* 6*d.*, or separately, sewed, 10*d.* each.

— Paradise Regained. Edited by K. DEIGHTON. 2*s.* 6*d.* ; sewed, 1*s.* 9*d.*

POPE, SELECTIONS FROM. Containing Essay on Criticism, Rape of the Lock, Temple of Fame, Windsor Forest. Edited by K. DEIGHTON. 2*s.* 6*d.* ; sewed, 1*s.* 9*d.*

SHAKESPEARE'S Julius Caesar. Edited by T. DUFF BARNETT, B.A. (Lond.). 2*s.*

— Merchant of Venice. Edited by T. DUFF BARNETT, B.A. (Lond.). 2*s.*

— Tempest. Edited by T. DUFF BARNETT, B.A. (Lond.). 2*s.*

Others to follow.

BELL'S READING BOOKS. Post 8vo, cloth, illustrated.

Infants.

Infant's Primer. 3*d.*
Tot and the Cat. 6*d.*
The Old Boathouse. 6*d.*
The Cat and the Hen. 6*d.*

Standard I.

School Primer. 6*d.*
The Two Parrots. 6*d.*
The Three Monkeys. 6*d.*
The New-born Lamb. 6*d.*
The Blind Boy. 6*d.*

Standard II.

The Lost Pigs. 6*d.*
Story of a Cat. 6*d.*
Queen Bee and Busy Bee. 6*d.*

Gulls' Crag. 6*d.*

Standard III.

Great Deeds in English History. 1*s.*
Adventures of a Donkey. 1*s.*
Grimm's Tales. 1*s.*
Great Englishmen. 1*s.*
Andersen's Tales. 1*s.*
Life of Columbus. 1*s.*

Standard IV.

Uncle Tom's Cabin. 1*s.*
Great Englishwomen. 1*s.*
Great Scotsmen. 1*s.*
Edgeworth's Tales. 1*s.*
Gatty's Parables from Nature. 1*s.*
Scott's Talisman. 1*s.*

BELL'S READING BOOKS—*continued.*

Standard V.

Dickens' Oliver Twist. 1s.
Dickens' Little Nell. 1s.
Masterman Ready. 1s.
Marryat's Poor Jack. 1s.
Arabian Nights. 1s.
Gulliver's Travels. 1s.
Lyrical Poetry for Boys and Girls. 1s.
Vicar of Wakefield. 1s.

Standards VI. and VII.

Lamb's Tales from Shakespeare. 1s.
Robinson Crusoe. 1s.
Tales of the Coast. 1s.
Settlers in Canada. 1s.
Southey's Life of Nelson. 1s.
Sir Roger de Coverley. 1s.

BELL'S GEOGRAPHICAL READERS. By M. J. BARRINGTON-WARD, M A. (Worcester College, Oxford).

The Child's Geography. Illustrated. Stiff paper cover, 6d.
The Map and the Compass. (Standard I) Illustrated. Cloth, 8d.

The Round World. (Standard II.) Illustrated. Cloth, 10d.
About England. (Standard III.) With Illustrations and Coloured Map. Cloth, 1s. 4d.

BELL'S ANIMAL LIFE READERS. A Series of Reading Books for the Standards, designed to inculcate the humane treatment of animals. Edited by EDITH CARRINGTON and ERNEST BELL. Illustrated by HARRISON WEIR and others. [*In preparation.*

EDWARDS (F.). Examples for Analysis in Verse and Prose. Selected and arranged by F. EDWARDS. *New edition.* Fcap. 8vo, cloth, 1s.

GOLDSMITH. The Deserted Village. Edited, with Notes and Life, by C. P. MASON, B.A., F.C.P. *4th edition.* Crown 8vo, 1s.

HANDBOOKS OF ENGLISH LITERATURE. Edited by J. W. HALES, M.A., formerly Clark Lecturer in English Literature at Trinity College, Cambridge, Professor of English Literature at King's College, London. Crown 8vo, 3s. 6d. each.

> The Age of Pope. By JOHN DENNIS.
> The Age of Dryden. By R. GARNETT, LL.D., C.B.
> *In preparation.*
> The Age of Chaucer. By PROFESSOR HALES.
> The Age of Shakespeare. By PROFESSOR HALES.
> The Age of Milton. By J. BASS MULLINGER, M.A.
> The Age of Wordsworth. By PROFESSOR C. H. HERFORD, LITT.D.
> The Age of Johnson. By THOMAS SECCOMBE.
> The Age of Tennyson. By PROFESSOR HUGH WALKER.

HAZLITT (W.). Lectures on the Literature of the Age of Elizabeth. Small post 8vo, sewed, 1s.

— Lectures on the English Poets. Small post 8vo, sewed, 1s.

— Lectures on the English Comic Writers. Small post 8vo, sewed, 1s.

LAMB (C.). Specimens of English Dramatic Poets of the Time of Elizabeth. With Notes. Small post 8vo, 3s. 6d.

MASON (C. P.). Grammars by C. P. MASON, B.A., F.C.P., Fellow of University College, London.

— First Notions of Grammar for Young Learners. Fcap. 8vo. *95th thousand.* Cloth, 1s.

— First Steps in English Grammar, for Junior Classes. Demy 18mo. *59th thousand.* 1s.

MASON (C. P.). Outlines of English Grammar, for the Use of Junior Classes. *17th edition. 97th thousand.* Crown 8vo, 2s.
— English Grammar; including the principles of Grammatical Analysis. *36th edition, revised. 153rd thousand.* Crown 8vo, green cloth, 3s. 6d.
— A Shorter English Grammar, with copious and carefully graduated Exercises, based upon the author's English Grammar. *9th edition. 49th thousand.* Crown 8vo, brown cloth, 3s. 6d.
— Practice and Help in the Analysis of Sentences. Price 2s. Cloth.
— English Grammar Practice, consisting of the Exercises of the Shorter English Grammar published in a separate form. *3rd edition.* Crown 8vo, 1s.
— Remarks on the Subjunctive and the so-called Potential Mood. 6d., sewn.
— Blank Sheets Ruled and headed for Analysis. 1s. per dozen.

MILTON : Paradise Lost. Books I., II., and III. Edited, with Notes on the Analysis and Parsing, and Explanatory Remarks, by C. P. MASON, B.A., F.C.P. Crown 8vo. 1s. each.
— Paradise Lost. Books V.-VIII. With Notes for the Use of Schools. By C. M. LUMBY. 2s. 6d.

PRICE (A. C.). Elements of Comparative Grammar and Philology. For Use in Schools. By A. C. PRICE, M.A., Assistant Master at Leeds Grammar School. Crown 8vo, 2s. 6d.

SHAKESPEARE. Notes on Shakespeare's Plays. With Introduction, Summary, Notes (Etymological and Explanatory), Prosody, Grammatical Peculiarities, etc. By T. DUFF BARNETT, B.A. Lond., late Second Master in the Brighton Grammar School. Specially adapted for the Local and Preliminary Examinations. Crown 8vo, 1s. each.
    Midsummer Night's Dream.—Julius Cæsar.—The Tempest.—Macbeth.—Henry V.—Hamlet.—Merchant of Venice.—King Richard II.—King John.—King Richard III.—King Lear.—Coriolanus.—Twelfth Night.—As You Like it.—Much Ado About Nothing.
    "The Notes are comprehensive and concise."—*Educational Times.*
    "Comprehensive, practical, and reliable."—*Schoolmaster.*
— Hints for Shakespeare-Study. Exemplified in an Analytical Study of Julius Cæsar. By MARY GRAFTON MOBERLY. *2nd edition.* Crown 8vo, sewed, 1s.
— Coleridge's Lectures and Notes on Shakespeare and other English Poets. Edited by T. ASHE, B.A. Small post 8vo, 3s. 6d.
— Shakespeare's Dramatic Art. The History and Character of Shakespeare's Plays. By DR. HERMANN ULRICI. Translated by L. DORA SCHMITZ. 2 vols. small post 8vo, 3s. 6d. each.
— William Shakespeare. A Literary Biography. By KARL ELZE, PH.D., LL.D. Translated by L. DORA SCHMITZ. Small post 8vo, 5s.
— Hazlitt's Lectures on the Characters of Shakespeare's Plays. Small post 8vo, 1s.
See BELL'S ENGLISH CLASSICS.
SKEAT (W. W.). Questions for Examinations in English Literature. With a Preface containing brief hints on the study of English. Arranged by the REV. W. W. SKEAT, LITT.D., Elrington and Bosworth

Professor of Anglo-Saxon in the University of Cambridge. *3rd edition.* Crown 8vo, 2*s.* 6*d.*

SMITH (C. J.) Synonyms and Antonyms of the English Language. Collected and Contrasted by the VEN. C. J. SMITH, M.A. *2nd edition,* *revised.* Small post 8vo, 5*s.*

— Synonyms Discriminated. A Dictionary of Synonymous Words in the English Language. Illustrated with Quotations from Standard Writers. By the late VEN. C. J. SMITH, M.A. With the Author's latest Corrections and Additions, edited by the REV. H. PERCY SMITH, M.A., of Balliol College, Oxford, Vicar of Great Barton, Suffolk. *4th edition.* Demy 8vo, 14*s.*

TEN BRINK'S History of English Literature. Vol. I. Early English Literature (to Wiclif). Translated into English by HORACE M. KENNEDY, Professor of German Literature in the Brooklyn Collegiate Institute. Small post 8vo, 3*s.* 6*d.*

— Vol. II. (Wiclif, Chaucer, Earliest Drama, Renaissance). Translated by W. CLARKE ROBINSON, PH.D. Small post 8vo, 3*s.* 6*d.*

— Lectures on Shakespeare. Translated by JULIA FRANKLIN. Small post 8vo, 3*s.* 6*d.*

THOMSON : Spring. Edited by C. P. MASON, B.A., F.C.P. With Life. *2nd edition.* Crown 8vo, 1*s.*

— Winter. Edited by C. P. MASON, B.A., F.C.P. With Life. Crown 8vo, 1*s.*

WEBSTER'S INTERNATIONAL DICTIONARY of the English Language. Including Scientific, Technical, and Biblical Words and Terms, with their Significations, Pronunciations, Alternative Spellings, Derivations, Synonyms, and numerous illustrative Quotations, with various valuable literary Appendices, with 83 extra pages of Illustrations grouped and classified, rendering the work a COMPLETE LITERARY AND SCIENTIFIC REFERENCE-BOOK. *New edition* (1890). Thoroughly revised and enlarged under the supervision of NOAH PORTER, D.D., LL.D. 1 vol. (2,118 pages, 3,500 woodcuts), 4to, cloth, 31*s.* 6*d.* ; half calf, £2 2*s.* ; half russia, £2 5*s.* ; calf, £2 8*s.* ; or in 2 vols. cloth, £1 14*s.*

Prospectuses, with specimen pages, sent post free on application.

WEBSTER'S BRIEF INTERNATIONAL DICTIONARY. A Pronouncing Dictionary of the English Language, abridged from Webster's International Dictionary. With a Treatise on Pronunciation, List of Prefixes and Suffixes, Rules for Spelling, a Pronouncing Vocabulary of Proper Names in History, Geography, and Mythology, and Tables of English and Indian Money, Weights, and Measures. With 564 pages and 800 Illustrations. Demy 8vo, 3*s.*

WRIGHT (T.). Dictionary of Obsolete and Provincial English. Containing Words from the English Writers previous to the 19th century, which are no longer in use, or are not used in the same sense, and Words which are now used only in the Provincial Dialects. Compiled by THOMAS WRIGHT, M.A., F.S.A., etc. 2 vols. 5*s.* each.

FRENCH CLASS BOOKS.

BOWER (A. M.). The Public Examination French Reader. With a Vocabulary to every extract, suitable for all Students who are preparing for a French Examination. By A. M. BOWER, F.R.G.S., late Master in University College School, etc. Cloth, 3*s.* 6*d.*

BARBIER (PAUL). A Graduated French Examination Course. By PAUL BARBIER, Lecturer in the South Wales University College, etc. Crown 8vo, 3*s.*

BARRERE (A.) Junior Graduated French Course. Affording Materials for Translation, Grammar, and Conversation. By A. BARRÈRE, Professor R.M.A., Woolwich. 1*s.* 6*d.*

— Elements of French Grammar and First Steps in Idioms. With numerous Exercises and a Vocabulary. Being an Introduction to the Précis of Comparative French Grammar. Crown 8vo, 2*s.*

— Précis of Comparative French Grammar and Idioms and Guide to Examinations. *4th edition.* 3*s.* 6*d.*

— Récits Militaires. From Valmy (1792) to the Siege of Paris (1870). With English Notes and Biographical Notices. *2nd edition.* Crown 8vo, 3*s.*

CLAPIN (A. C.). French Grammar for Public Schools. By the REV. A. C. CLAPIN, M.A., St. John's College, Cambridge, and Bachelier-ès-lettres of the University of France. Fcap. 8vo. *14th edition.* 2*s.* 6*d.* Key to the Exercises. 3*s.* 6*d.* net.

— French Primer. Elementary French Grammar and Exercises for Junior Forms in Public and Preparatory Schools. Fcap. 8vo. *10th edition.* 1*s.*

— Primer of French Philology. With Exercises for Public Schools. *7th edition.* Fcap. 8vo, 1*s.*

— English Passages for Translation into French. Crown 8vo, 2*s.* 6*d.* Key (for Tutors only), 4*s.* net.

DAVIS (J. F.) Army Examination Papers in French. Questions set at the Preliminary Examinations for Sandhurst and Woolwich, from Nov., 1876, to June, 1890, with Vocabulary. By J. F. DAVIS, D.LIT., M.A., Lond. Crown 8vo, 2*s.* 6*d.*

DAVIS (J. F.) and THOMAS (F.). An Elementary French Reader. Compiled, with a Vocabulary, by J. F. DAVIS, M.A., D.LIT., and FERDINAND THOMAS, Assistant Examiners in the University of London. Crown 8vo, 2*s.*

DELILLE'S GRADUATED FRENCH COURSE.

The Beginner's own French Book. 2*s.* Key, 2*s.*	Repertoire des Prosateurs. 3*s.* 6*d.* Modèles de Poesie. 3*s.* 6*d.*
Easy French Poetry for Beginners. 2*s.*	Manuel Etymologique. 2*s.* 6*d.*
French Grammar. 3*s.* Key, 3*s.*	Synoptical Table of French Verbs. 6*d.*

ESCLANGON (A.). The French Verb Newly Treated : an Easy, Uniform, and Synthetic Method of its Conjugation. By A. ESCLANGON, Examiner in the University of London. Small 4to, 5*s.*

GASC (F. E. A.). First French Book ; being a New, Practical, and Easy Method of Learning the Elements of the French Language. *Reset and thoroughly revised.* 116*th thousand.* Crown 8vo, 1*s.*

— Second French Book ; being a Grammar and Exercise Book, on a new and practical plan, and intended as a sequel to the " First French Book." 52*nd thousand.* Fcap. 8vo, 1*s.* 6*d.*

GASC (F. E. A.). Key to First and Second French Books. *6th edition,* Fcap. 8vo, 3s. 6d. net.

— French Fables, for Beginners, in Prose, with an Index of all the Words at the end of the work. *17th thousand.* 12mo, 1s. 6d.

— Select Fables of La Fontaine. *19th thousand.* Fcap. 8vo, 1s. 6d.

— Histoires Amusantes et Instructives; or, Selections of Complete Stories from the best French modern authors, who have written for the young. With English notes. *17th thousand.* Fcap. 8vo, 2s.

— Practical Guide to Modern French Conversation, containing:— I. The most current and useful Phrases in Everyday Talk. II. Everybody's necessary Questions and Answers in Travel-Talk. *19th edition.* Fcap. 8vo, 1s. 6d.

— French Poetry for the Young. With Notes, and preceded by a few plain Rules of French Prosody. *5th edition, revised.* Fcap. 8vo, 1s. 6d.

— French Prose Composition, Materials for. With copious footnotes, and hints for idiomatic renderings. *21st thousand.* Fcap. 8vo, 3s.
Key. *2nd edition.* 6s. net.

— Prosateurs Contemporains; or, Selections in Prose chiefly from contemporary French literature. With notes. *11th edition.* 12mo, 3s. 6d.

— Le Petit Compagnon; a French Talk-Book for Little Children. *14th edition.* 16mo, 1s. 6d.

— French and English Dictionary, with upwards of Fifteen Thousand new words, senses, &c., hitherto unpublished. *5th edition, with numerous additions and corrections.* In one vol. 8vo, cloth, 10s. 6d. In use at Harrow, Rugby, Shrewsbury, &c.

— Pocket Dictionary of the French and English Languages; for the everyday purposes of Travellers and Students. Containing more than Five Thousand modern and current words, senses, and idiomatic phrases and renderings, not found in any other dictionary of the two languages. *New edition. 53rd thousand.* 16mo, cloth, 2s. 6d.

GOSSET (A.). Manual of French Prosody for the use of English Students. By ARTHUR GOSSET, M.A., Fellow of New College, Oxford. Crown 8vo, 3s.
"This is the very book we have been looking for. We hailed the title with delight, and were not disappointed by the perusal. The reader who has mastered the contents will know, what not one in a thousand of Englishmen who read French knows, the rules of French poetry."— *Journal of Education.*

LE NOUVEAU TRESOR; designed to facilitate the Translation of English into French at Sight. By M. E. S. *18th edition.* Fcap. 8vo, 1s. 6d.

STEDMAN (A. M. M.). French Examination Papers in Miscellaneous Grammar and Idioms. Compiled by A. M. M. STEDMAN, M.A. *5th edition.* Crown 8vo, 2s. 6d.
A Key. By G. A. SCHRUMPF. For Tutors only. 6s. net.

— Easy French Passages for Unseen Translation. Fcap. 8vo, 1s. 6d.

— Easy French Exercises on Elementary Syntax. Crown 8vo, 2s. 6d.

— First French Lessons. Crown 8vo, 1s.

— French Vocabularies for Repetition. Fcap. 8vo, 1s.

— Steps to French. 12mo, 8d.

FRENCH ANNOTATED EDITIONS.

BALZAC. Ursule Mirouët. By HONORÉ DE BALZAC. Edited, with Introduction and Notes, by JAMES BOÏELLE, B.-ès-L., Senior French Master, Dulwich College. 3*s*.

CLARÉTIE. Pierrille. By JULES CLARÉTIE. With 27 Illustrations. Edited, with Introduction and Notes, by JAMES BOÏELLE, B.-ès-L. 2*s*. 6*d*.

DAUDET. La Belle Nivernaise. Histoire d'un vieux bateau et de son équipage. By ALPHONSE DAUDET. Edited, with Introduction and Notes, by JAMES BOÏELLE, B.-ès-L. With Six Illustrations. 2*s*.

FÉNELON. Aventures de Télémaque. Edited by C. J. DELILLE. *4th edition.* Fcap. 8vo, 2*s*. 6*d*.

GOMBERT'S FRENCH DRAMA. Re-edited, with Notes, by F. E. A. GASC. Sewed, 6*d*. each.

MOLIÈRE.

Le Misanthrope.
L'Avare.
Le Bourgeois Gentilhomme.
Le Tartuffe.
Le Malade Imaginaire.
Les Femmes Savantes.

Les Fourberies de Scapin.
Les Précieuses Ridicules.
L'Ecole des Femmes.
L'Ecole des Maris.
Le Médecin Malgré Lui.

RACINE.

La Thébaïde, ou Les Frères Ennemis.
Andromaque.
Les Plaideurs.
Iphigénie.

Britannicus.
Phèdre.
Esther.
Athalie.

CORNEILLE.

Le Cid.
Horace.

Cinna.
Polyeucte.

VOLTAIRE.—Zaïre.

GREVILLE. Le Moulin Frappier. By HENRY GREVILLE. Edited, with Introduction and Notes, by JAMES BOÏELLE, B.-ès-L. 3*s*.

HUGO. Bug Jargal. Edited, with Introduction and Notes, by JAMES BOÏELLE, B.-ès-L. 3*s*.

LA FONTAINE. Select Fables. Edited by F. E. A. GASC. 19*th thousand*. Fcap. 8vo, 1*s*. 6*d*.

LAMARTINE. Le Tailleur de Pierres de Saint-Point. Edited with Notes by JAMES BOÏELLE, B.-ès-L. *6th thousand.* Fcap. 8vo, 1*s*. 6*d*.

SAINTINE. Picciola. Edited by DR. DUBUC. 16*th thousand*. Fcap. 8vo, 1*s*. 6*d*.

VOLTAIRE. Charles XII. Edited by L. DIREY. *7th edition.* Fcap. 8vo, 1*s*. 6*d*.

GERMAN CLASS BOOKS.

BUCHHEIM (DR. C. A.). German Prose Composition. Consisting of Selections from Modern English Writers. With grammatical notes, idiomatic renderings, and general introduction. By C. A. BUCHHEIM, PH.D., Professor of the German Language and Literature in King's College, and Examiner in German to the London University. 14*th edition, enlarged and revised.* With a list of subjects for original composition. Fcap. 8vo, 4*s*. 6*d*.

A KEY to the 1st and 2nd parts. *3rd edition.* 3*s.* net. To the 3rd and 4th parts. 4*s.* net.

BUCHHEIM (DR. C. A.). First Book of German Prose. Being Parts I. and II. of the above. With Vocabulary by H. R. Fcap. 8vo, 1*s.* 6*d.*

CLAPIN (A. C.). A German Grammar for Public Schools. By the REV. A. C. CLAPIN, and F. HOLL-MÜLLER, Assistant Master at the Bruton Grammar School. *6th edition.* Fcap. 8vo, 2*s.* 6*d.*

— A German Primer. With Exercises. *2nd edition.* Fcap. 8vo, 1*s.*

German. The Candidate's Vade Mecum. Five Hundred Easy Sentences and Idioms. By an Army Tutor. Cloth, 1*s.* For Army Prelim. Exam.

LANGE (F.). A Complete German Course for Use in Public Schools. By F. LANGE, PH.D., Professor R.M.A. Woolwich, Examiner in German to the College of Preceptors, London ; Examiner in German at the Victoria University, Manchester. Crown 8vo.

Concise German Grammar. With special reference to Phonology, Comparative Philology, English and German Equivalents and Idioms. Comprising Materials for Translation, Grammar, and Conversation. Elementary, 2*s.* ; Intermediate, 2*s.* ; Advanced, 3*s.* 6*d.*

Progressive German Examination Course. Comprising the Elements of German Grammar, an Historic Sketch of the Teutonic Languages, English and German Equivalents, Materials for Translation, Dictation, Extempore Conversation, and Complete Vocabularies. I. Elementary Course, 2*s.* II. Intermediate Course, 2*s.* III. Advanced Course. *Second revised edition.* 1*s.* 6*d.*

Elementary German Reader. A Graduated Collection of Readings in Prose and Poetry. With English Notes and a Vocabulary. *4th edition.* 1*s.* 6*d.*

Advanced German Reader. A Graduated Collection of Readings in Prose and Poetry. With English Notes by F. LANGE, PH.D., and J. F. DAVIS, D.LIT. *2nd edition.* 3*s.*

MORICH (R. J.). German Examination Papers in Miscellaneous Grammar and Idioms. By R. J. MORICH, Manchester Grammar School. *2nd edition.* Crown 8vo, 2*s.* 6*d.* A Key, for Tutors only. 5*s.* net.

PHILLIPS (M. E.). Handbook of German Literature. By MARY E. PHILLIPS, LL.A. With Introduction by DR. A. WEISS, Professor of German Literature at R. M. A. Woolwich. Crown 8vo. [*Shortly.*

STOCK (DR.). Wortfolge, or Rules and Exercises on the order of Words in German Sentences. With a Vocabulary. By the late FREDERICK STOCK, D.LIT., M.A. Fcap. 8vo, 1*s.* 6*d.*

KLUGE'S Etymological Dictionary of the German Language. Translated by J. F. DAVIS, D.LIT. (Lond.). Crown 4to, 18*s.*

GERMAN ANNOTATED EDITIONS.

AUERBACH (B.). Auf Wache. Novelle von BERTHOLD AUERBACH. Der Gefrorene Kuss. Novelle von OTTO ROQUETTE. Edited by A. A. MACDONELL, M.A., PH.D. *2nd edition.* Crown 8vo, 2*s.*

BENEDIX (J. R.). Doktor Wespe. Lustspiel in fünf Aufzügen von JULIUS RODERICH BENEDIX. Edited by PROFESSOR F. LANGE, PH.D. Crown 8vo, 2*s.* 6*d.*

EBERS (G.). Eine Frage. Idyll von GEORG EBERS. Edited by F. STORR. B.A., Chief Master of Modern Subjects in Merchant Taylors' School. Crown 8vo, 2s.

FREYTAG (G.). Die Journalisten. Lustspiel von GUSTAV FREYTAG. Edited by PROFESSOR F. LANGE, PH.D. *4th revised edition.* Crown 8vo, 2s. 6d.

— **SOLL UND HABEN.** Roman von GUSTAV FREYTAG. Edited by W. HANBY CRUMP, M.A. Crown 8vo, 2s. 6d.

GERMAN BALLADS from Uhland, Goethe, and Schiller. With Introductions, Copious and Biographical Notices. Edited by C. L. BIELEFELD. *4th edition.* Fcap. 8vo, 1s. 6d.

GERMAN EPIC TALES IN PROSE. I. Die Nibelungen, von A. F. C. VILMAR. II. Walther und Hildegund, von ALBERT RICHTER. Edited by KARL NEUHAUS, PH.D., the International College, Isleworth. Crown 8vo, 2s. 6d.

GOETHE. Hermann und Dorothea. With Introduction, Notes, and Arguments. By E. BELL, M.A., and E. WÖLFEL. *2nd edition.* Fcap. 8vo, 1s. 6d.

GOETHE. FAUST. Part I. German Text with Hayward's Prose Translation and Notes. Revised, With Introduction by C. A. BUCHHEIM, PH.D., Professor of German Language and Literature at King's College, London. Small post 8vo, 5s.

GUTZKOW (K.). Zopf und Schwert. Lustspiel von KARL GUTZKOW. Edited by PROFESSOR F. LANGE, PH.D. Crown 8vo, 2s. 6d.

HEY'S FABELN FÜR KINDER. Illustrated by O. SPECKTER. Edited, with an Introduction, Grammatical Summary, Words, and a complete Vocabulary, by PROFESSOR F. LANGE, PH.D. Crown 8vo, 1s. 6d.

— The same. With a Phonetic Introduction, and Phonetic Transcription of the Text. By PROFESSOR F. LANGE, PH.D. Crown 8vo, 2s.

HEYSE (P.). Hans Lange. Schauspiel von PAUL HEYSE. Edited by A. A. MACDONELL, M.A., PH.D., Taylorian Teacher, Oxford University. Crown 8vo, 2s.

HOFFMANN (E. T. A.). Meister Martin, der Küfner. Erzählung von E. T. A. HOFFMANN. Edited by F. LANGE, PH.D. *2nd edition.* Crown 8vo, 1s. 6d.

MOSER (G. VON). Der Bibliothekar. Lustspiel von G. VON MOSER. Edited by F. LANGE, PH.D. *4th edition.* Crown 8vo, 2s.

ROQUETTE (O.). See Auerbach.

SCHEFFEL (V. VON). Ekkehard. Erzählung des zehnten Jahrhunderts, von VICTOR VON SCHEFFEL. Abridged edition, with Introduction and Notes by HERMAN HAGER, PH.D., Lecturer in the German Language and Literature in The Owens College, Victoria University, Manchester. Crown 8vo, 3s.

SCHILLER'S Wallenstein. Complete Text, comprising the Weimar Prologue, Lager, Piccolomini, and Wallenstein's Tod. Edited by DR. BUCHHEIM, Professor of German in King's College, London. *6th edition.* Fcap. 8vo, 5s. Or the Lager and Piccolomini, 2s. 6d. Wallenstein's Tod, 2s. 6d.

— Maid of Orleans. With English Notes by DR. WILHELM WAGNER. *3rd edition.* Fcap. 8vo, 1s. 6d.

— Maria Stuart. Edited by V. KASTNER, B.-ès-L., Lecturer on French Language and Literature at Victoria University, Manchester. *3rd edition.* Fcap. 8vo, 1s. 6d.

ITALIAN.

DANTE. The Inferno. A Literal Prose Translation, with the Text of the Original collated with the best editions, printed on the same page, and Explanatory Notes. By JOHN A. CARLYLE, M.D. With Portrait. *2nd edition.* Small post 8vo, 5*s.*

— The Purgatorio. A Literal Prose Translation, with the Text of Bianchi printed on the same page, and Explanatory Notes. By W. S. DUGDALE. Small post 8vo, 5*s.*

BELL'S MODERN TRANSLATIONS.

A Series of Translations from Modern Languages, with Memoirs, Introductions, etc. Crown 8vo, 1*s.* each.

GOETHE. Egmont. Translated by ANNA SWANWICK.
— Iphigenia in Tauris. Translated by ANNA SWANWICK.
HAUFF. The Caravan. Translated by S. MENDEL.
— The Inn in the Spessart. Translated by S. MENDEL.
LESSING. Laokoon. Translated by E. C. BEASLEY.
— Nathan the Wise. Translated by R. DILLON BOYLAN.
— Minna von Barnhelm. Translated by ERNEST BELL, M.A.
MOLIÈRE. The Misanthrope. Translated by C. HERON WALL.
— The Doctor in Spite of Himself. (Le Médecin malgré lui). Translated by C. HERON WALL.
— Tartuffe; or, The Impostor. Translated by C. HERON WALL.
— The Miser. (L'Avare). Translated by C. HERON WALL.
— The Shopkeeper turned Gentleman. (Le Bourgeois Gentilhomme). Translated by C. HERON WALL.
RACINE. Athalie. Translated by R. BRUCE BOSWELL, M.A.
— Esther. Translated by R. BRUCE BOSWELL, M.A.
SCHILLER. William Tell. Translated by SIR THEODORE MARTIN, K.C.B., LL.D. *New edition, entirely revised.*
— The Maid of Orleans. Translated by ANNA SWANWICK.
— Mary Stuart. Translated by J. MELLISH.
— Wallenstein's Camp and the Piccolomini. Translated by J. CHURCHILL and S. T. COLERIDGE.
— The Death of Wallenstein. Translated by S. T. COLERIDGE.

*** For other Translations of Modern Languages, *see* the Catalogue of Bohn's Libraries, which will be forwarded on application.

SCIENCE, TECHNOLOGY, AND ART.

CHEMISTRY.

COOKE (S.). First Principles of Chemistry. An Introduction to Modern Chemistry for Schools and Colleges. By SAMUEL COOKE, M.A., B.E., Assoc. Mem. Inst. C. E., Principal of the College of Science, Poona. *6th edition, revised.* Crown 8vo, 2*s.* 6*d.*

— The Student's Practical Chemistry. Test Tables for Qualitative Analysis. *3rd edition, revised and enlarged.* Demy 8vo, 1*s.*

STÖCKHARDT (J. A.). **Experimental Chemistry.** Founded on the work of J. A. STÖCKHARDT. A Handbook for the Study of Science by Simple Experiments. By C. W. HEATON, F.I.C., F.C.S., Lecturer in Chemistry in the Medical School of Charing Cross Hospital, Examiner in Chemistry to the Royal College of Physicians, etc. *Revised edition.* 5*s.*

WILLIAMS (W. M.). **The Framework of Chemistry.** Part I. Typical Facts and Elementary Theory. By W. M. WILLIAMS, M.A., St. John's College, Oxford ; Science Master, King Henry VIII.'s School, Coventry. Crown 8vo, paper boards, 9*d.* net.

BOTANY.

HAYWARD (W. R.). **The Botanist's Pocket-Book.** Containing in a tabulated form, the chief characteristics of British Plants, with the botanical names, soil, or situation, colour, growth, and time of flowering of every plant, arranged under its own order ; with a copious Index. By W. R. HAYWARD. *6th edition, revised.* Fcap. 8vo, cloth limp, 4*s.* 6*d.*

LONDON CATALOGUE of British Plants. Part I., containing the British Phænogamia, Filices, Equisetaceæ, Lycopodiaceæ, Selaginellaceæ, Marsileaceæ, and Characeæ. *9th edition.* Demy 8vo, 6*d.* ; interleaved in limp cloth, 1*s.* Generic Index only, on card, 2*d.*

MASSEE (G.). **British Fungus-Flora.** A Classified Text-Book of Mycology. By GEORGE MASSEE, Author of "The Plant World." With numerous Illustrations. 4 vols. post 8vo, 7*s.* 6*d.* each.

SOWERBY'S English Botany. Containing a Description and Life-size Drawing of every British Plant. Edited and brought up to the present standard of scientific knowledge, by T. BOSWELL (late SYME), LL.D., F.L.S., etc. *3rd edition, entirely revised.* With Descriptions of all the Species by the Editor, assisted by N. E. BROWN. 12 vols., with 1,937 *coloured plates,* £24 3*s.* in cloth, £26 11*s.* in half-morocco, and £30 9*s.* in whole morocco. Also in 89 parts, 5*s.*, except Part 89, containing an Index to the whole work, 7*s.* 6*d.*

.*. A Supplement, to be completed in 8 or 9 parts, is now publishing. Parts I., II., and III. ready, 5*s.* each, or bound together, making Vol. XIII. of the complete work, 17*s.*

TURNBULL (R.). **Index of British Plants,** according to the London Catalogue (Eighth Edition), including the Synonyms used by the principal authors, an Alphabetical List of English Names, etc. By ROBERT TURNBULL. Paper cover, 2*s.* 6*d.*, cloth, 3*s.*

GEOLOGY.

JUKES-BROWNE (A. J.). **Student's Handbook of Physical Geo-logy.** By A. J. JUKES-BROWNE, B.A., F.G.S., of the Geological Survey of England and Wales. With numerous Diagrams and Illustrations. *2nd edition, much enlarged,* 7*s.* 6*d.*

— Student's Handbook of Historical Geology. With numerous Diagrams and Illustrations. 6*s.*

"An admirably planned and well executed ' Handbook of Historical Geology.'"—*Journal of Education.*

— The Building of the British Isles. A Study in Geographical Evolution With Maps. *2nd edition revised.* 7*s.* 6*d.*

MEDICINE.

CARRINGTON (R. E.), and LANE (W. A.). A Manual of Dissec-
tions of the Human Body. By the late R. E. CARRINGTON, M.D.
(Lond.), F.R.C.P., Senior Assistant Physician, Guy's Hospital. *2nd
edition.* Revised and enlarged by W. ARBUTHNOT LANE, M.S., F.R.C.S.,
Assistant Surgeon to Guy's Hospital, etc. Crown 8vo, 9s.
   " As solid a piece of work as ever was put into a book ; accurate from
   beginning to end, and unique of its kind."—*British Medical Journal.*
HILTON'S Rest and Pain. Lectures on the Influence of Mechanical and
Physiological Rest in the Treatment of Accidents and Surgical Diseases,
and the Diagnostic Value of Pain. By the late JOHN HILTON, F.R.S.,
F.R.C.S., etc. Edited by W. H. A. JACOBSON, M.A., M.CH. (Oxon.),
F.R.C.S. *5th edition.* 9s.
**HOBLYN'S Dictionary of Terms used in Medicine and the Collateral
Sciences.** *12th edition.* Revised and enlarged by J. A. P. PRICE, B.A.,
M.D. (Oxon.). 10s. 6d.
LANE (W. A.). Manual of Operative Surgery. For Practitioners and
Students. By W. ARBUTHNOT LANE, M.B., M.S., F.R.C.S., Assistant
Surgeon to Guy's Hospital. Crown 8vo, 8s. 6d.
SHARP (W.) Therapeutics founded on Antipraxy. By WILLIAM
SHARP, M.D., F.R.S. Demy 8vo, 6s.

BELL'S AGRICULTURAL SERIES.

In crown 8vo, Illustrated, 160 pages, cloth, 2s. 6d. each.

CHEAL (J.). Fruit Culture. A Treatise on Planting, Growing, Storage
of Hardy Fruits for Market and Private Growers. By J. CHEAL, F.R.H.S.,
Member of Fruit Committee, Royal Hort. Society, etc.
FREAM (DR.). Soils and their Properties. By DR. WILLIAM FREAM,
B.SC. (Lond.)., F.L.S., F.G.S., F.S.S., Associate of the Surveyor's Institu-
tion, Consulting Botanist to the British Dairy Farmers' Association and
the Royal Counties Agricultural Society ; Prof. of Nat. Hist. in Downton
College, and formerly in the Royal Agric. Coll., Cirencester.
GRIFFITHS (DR.). Manures and their Uses. By DR. A. B. GRIFFITHS,
F.R.S.E., F.C.S., late Principal of the School of Science, Lincoln ; Membre
de la Société Chimique de Paris ; Author of " A Treatise on Manures,"
etc., etc. *In use at Downton College.*
— **The Diseases of Crops and their Remedies.**
MALDEN (W. J.). Tillage and Implements. By W. J. MALDEN,
Prof. of Agriculture in the College, Downton.
SHELDON (PROF.). The Farm and the Dairy. By PROFESSOR
J. P. SHELDON, formerly of the Royal Agricultural College, and of the
Downton College of Agriculture, late Special Commissioner of the
Canadian Government. *In use at Downton College.*

Specially adapted for Agricultural Classes. Crown 8vo. Illustrated. 1s. each.
Practical Dairy Farming. By PROFESSOR SHELDON. Reprinted from the
author's larger work entitled " The Farm and the Dairy."
Practical Fruit Growing. By J. CHEAL, F.R.H.S. Reprinted from the
author's larger work, entitled " Fruit Culture."

TECHNOLOGICAL HANDBOOKS.

Edited by Sir H. Trueman Wood.

Specially adapted for candidates in the examinations of the City Guilds Institute. Illustrated and uniformly printed in small post 8vo.

BEAUMONT (R.). Woollen and Worsted Cloth Manufacture. By ROBERTS BEAUMONT, Professor of Textile Industry, Yorkshire College, Leeds; Examiner in Cloth Weaving to the City and Guilds of London Institute. *2nd edition.* 7s. 6d.

BENEDIKT (R), and KNECHT (E.). Coal-tar Colours, The Chemistry of. With special reference to their application to Dyeing, etc. By DR. R. BENEDIKT, Professor of Chemistry in the University of Vienna. Translated by E. KNECHT, PH.D. of the Technical College, Bradford. *2nd and enlarged edition,* 6s. 6d.

CROOKES (W.). Dyeing and Tissue-Printing. By WILLIAM CROOKES, F.R.S., V.P.C.S. 5s.

GADD (W. L.). Soap Manufacture. By W. LAWRENCE GADD, F.I.C., F.C.S., Registered Lecturer on Soap-Making and the Technology of Oils and Fats, also on Bleaching, Dyeing, and Calico Printing, to the City and Guilds of London Institute. 5s.

HELLYER (S. S.). Plumbing: Its Principles and Practice. By S. STEVENS HELLYER. With numerous Illustrations. 5s.

HORNBY (J.). Gas Manufacture. By J. HORNBY, F.I.C., Lecturer under the City and Guilds of London Institute. [*In the press.*

HURST (G.H.). Silk-Dyeing and Finishing. By G. H. HURST, F.C.S., Lecturer at the Manchester Technical School, Silver Medallist, City and Guilds of London Institute. With Illustrations and numerous Coloured Patterns. 7s. 6d.

JACOBI (C. T.). Printing. A Practical Treatise. By C. T. JACOBI, Manager of the Chiswick Press, Examiner in Typography to the City and Guilds of London Institute. With numerous Illustrations. 5s.

MARSDEN (R.). Cotton Spinning: Its Development, Principles, and Practice, with Appendix on Steam Boilers and Engines. By R. MARSDEN, Editor of the "Textile Manufacturer." *4th edition.* 6s. 6d.

— Cotton Weaving: Its Development, Principles, and Practice. By R. MARSDEN. With numerous Illustrations. 10s. 6d.

PHILLIPSON (J.). Coach Building. [*Preparing.*

POWELL (H.), CHANCE (H.), and HARRIS (H. G.). Glass Manufacture. Introductory Essay, by H. POWELL, B.A. (Whitefriars Glass Works) ; Sheet Glass, by HENRY CHANCE, M.A. (Chance Bros., Birmingham): Plate Glass, by H. G. HARRIS, Assoc. Memb. Inst. C.E. 3s. 6d.

ZAEHNSDORF (J. W.) Bookbinding. By J. W. ZAEHNSDORF, Examiner in Bookbinding to the City and Guilds of London Institute. With 8 Coloured Plates and numerous Diagrams. *2nd edition, revised and enlarged.* 5s.

*** *Complete List of Technical Books on Application.*

MUSIC.

BANISTER (H. C.). A Text Book of Music : By H. C. BANISTER, Professor of Harmony and Composition at the R.A. of Music, at the Guild-hall School of Music, and at the Royal Normal Coll. and Acad. of Music for the Blind. 15th edition. Fcap. 8vo. 5s.

This Manual contains chapters on Notation, Harmony, and Counterpoint ;

BANISTER (H. C.)—*continued.*
Modulation, Rhythm, Canon, Fugue, Voices, and Instruments; together with exercises on Harmony, an Appendix of Examination Papers, and a copious Index and Glossary of Musical Terms.

— **Lectures on Musical Analysis.** Embracing Sonata Form, Fugue, etc., Illustrated by the Works of the Classical Masters. *2nd edition, revised.* Crown 8vo, 7s. 6d.

— **Musical Art and Study** : Papers for Musicians. Fcap. 8vo, 2s.

CHATER (THOMAS). Scientific Voice, Artistic Singing, and Effective Speaking. A Treatise on the Organs of the Voice, their Natural Functions, Scientific Development, Proper Training, and Artistic Use. By THOMAS CHATER. With Diagrams. Wide fcap. 2s. 6d.

HUNT (H. G. BONAVIA). A Concise History of Music, from the Commencement of the Christian era to the present time. For the use of Students. By REV. H. G. BONAVIA HUNT, Mus. Doc. Dublin ; Warden of Trinity College, London ; and Lecturer on Musical History in the same College. 13th edition, revised to date (1895). Fcap. 8vo, 3s. 6d.

ART.

BARTER (S.) Manual Instruction—Woodwork. By S. BARTER Organizer and Instructor for the London School Board, and to the Joint Committee on Manual Training of the School Board for London, the City and Guilds of London Institute, and the Worshipful Company of Drapers. With over 300 Illustrations. Fcap. 4to, cloth. 7s. 6d.

BELL (SIR CHARLES). The Anatomy and Philosophy of Expression, as connected with the Fine Arts. By SIR CHARLES BELL, K.H. *7th edition, revised.* 5s.

BRYAN'S Biographical and Critical Dictionary of Painters and Engravers. With a List of Ciphers, Monograms, and Marks. A new Edition, thoroughly Revised and Enlarged. By R. E. GRAVES and WALTER ARMSTRONG. 2 volumes. Imp. 8vo, buckram, 3l. 3s.

CHEVREUL on Colour. Containing the Principles of Harmony and Contrast of Colours, and their Application to the Arts. *3rd edition,* with Introduction. Index and several Plates. 5s.—With an additional series of 16 Plates in Colours, 7s. 6d.

DELAMOTTE (P. H.). The Art of Sketching from Nature. By P. H. DELAMOTTE, Professor of Drawing at King's College, London. Illustrated by Twenty-four Woodcuts and Twenty Coloured Plates, arranged progressively, from Water-colour Drawings by PROUT, E. W. COOKE, R.A., GIRTIN, VARLEY, DE WINT, and the Author. *New edition.* Imp. 4to, 21s.

FLAXMAN'S CLASSICAL COMPOSITIONS, reprinted in a cheap form for the use of Art Students. Oblong paper covers, 2s. 6d. each.
Homer. 2 vols.—Æschylus.—Hesiod.—Dante.

— **Lectures on Sculpture,** as delivered before the President and Members of the Royal Academy. With Portrait and 53 plates. 6s.

HARRIS (R.). Geometrical Drawing. For Army and other Examinations. With chapters on Scales and Graphic Statics. With 221 diagrams. By R. HARRIS, Art Master at St. Paul's School. *New edition, enlarged.* Crown 8vo, 3s. 6d.

HEATON (MRS.). A Concise History of Painting. By the late MRS. CHARLES HEATON. *New edition.* Revised by COSMO MONKHOUSE. 5s.

LELAND (C. G.). Drawing and Designing. In a series of Lessons for School use and Self Instruction. By CHARLES G. LELAND, M.A., F.R.L.S. Paper cover, 1s. ; or in cloth, 1s. 6d.
— Leather Work: Stamped, Moulded, and Cut, Cuir-Bouillé, Sewn, etc. With numerous Illustrations. Fcap. 4to, 5s.
— Manual of Wood Carving. By CHARLES G. LELAND, M.A., F.R.L.S. Revised by J. J. HOLTZAPFFEL, A.M. INST.C.E. With numerous Illustrations. Fcap. 4to, 5s.
— Metal Work. With numerous Illustrations. Fcap. 4to, 5s.
LEONARDO DA VINCI'S Treatise on Painting. Translated from the Italian by J. F. RIGAUD, R.A. With a Life of Leonardo and an Account of his Works, by J. W. BROWN. With numerous Plates. 5s.
MOODY (F. W.). Lectures and Lessons on Art. By the late F. W. MOODY, Instructor in Decorative Art at South Kensington Museum. With Diagrams to illustrate Composition and other matters. *A new and cheaper edition.* Demy 8vo, sewed, 4s. 6d.
STRANGE (E. F). Alphabets : a Handbook of Lettering, compiled for the use of Artists, Designers, Handicraftsmen, and Students. With complete Historical and Practical Descriptions. By EDWARD F. STRANGE. With more than 200 Illustrations. Imperial 16mo, 8s. 6d. net.
WHITE (GLEESON). Practical Designing: A Handbook on the Preparation of Working Drawings, showing the Technical Methods employed in preparing them for the Manufacturer and the Limits imposed on the Design by the Mechanism of Reproduction and the Materials employed. Edited by GLEESON WHITE. Freely Illustrated. *2nd edition.* Crown 8vo, 6s. net.
Contents :—Bookbinding, by H. ORRINSMITH—Carpets, by ALEXANDER MILLAR—Drawing for Reproduction, by the Editor—Pottery, by W. P. RIX—Metal Work, by R. LL. RATHBONE—Stained Glass, by SELWYN IMAGE—Tiles, by OWEN CARTER—Woven Fabrics, Printed Fabrics, and Floorcloths, by ARTHUR SILVER—Wall Papers, by G. C. HAITÉ.

MENTAL, MORAL, AND SOCIAL SCIENCES.

PSYCHOLOGY AND ETHICS.

ANTONINUS (M. Aurelius). The Thoughts of. Translated literally, with Notes, Biographical Sketch, Introductory Essay on the Philosophy, and Index, by GEORGE LONG, M.A. *Revised edition.* Small post 8vo, 3s. 6d., *or new edition on Handmade paper, buckram,* 6s.
BACON'S Novum Organum and Advancement of Learning. Edited, with Notes, by J. DEVEY, M.A. Small post 8vo, 5s.
EPICTETUS. The Discourses of. With the Encheiridion and Fragments. Translated with Notes, a Life of Epictetus, a View of his Philosophy, and Index, by GEORGE LONG, M.A. Small post 8vo, 5s., *or new edition on Handmade paper,* 2 vols., *buckram,* 10s. 6d.
KANT'S Critique of Pure Reason. Translated by J. M. D. MEIKLEJOHN, Professor of Education at St. Andrew's University. Small post 8vo, 5s.
— Prolegomena and Metaphysical Foundations of Science. With Life. Translated by E. BELFORT BAX. Small post 8vo, 5s.
LOCKE'S Philosophical Works. Edited by J. A. ST. JOHN. 2 vols. Small post 8vo, 3s. 6d. each.

RYLAND (F.). The Student's Manual of Psychology and Ethics, designed chiefly for the London B.A. and B.Sc. By F. RYLAND, M.A., late Scholar of St. John's College, Cambridge. Cloth, red edges. *5th edition, revised and enlarged.* With lists of books for Students, and Examination Papers set at London University. Crown 8vo, 3*s.* 6*d.* .

— Ethics: An Introductory Manual for the use of University Students. With an Appendix containing List of Books recommended, and Examination Questions. Crown 8vo, 3*s.* 6*d.*

— Logic. An Introductory Manual. Crown 8vo. [*In the press.*

SCHOPENHAUER on the Fourfold Root of the Principle of Sufficient Reason, and On the Will in Nature. Translated by MADAME HILLEBRAND. Small post 8vo, 5*s.*

— Essays. Selected and Translated. With a Biographical Introduction and Sketch of his Philosophy, by E. BELFORT BAX. Small post 8vo, 5*s.*

SMITH (Adam). Theory of Moral Sentiments. With Memoir of the Author by DUGALD STEWART. Small post 8vo, 3*s.* 6*d.*

SPINOZA'S Chief Works. Translated with Introduction, by R. H. M. ELWES. 2 vols. Small post 8vo, 5*s.* each.
> Vol. I.—Tractatus Theologico-Politicus—Political Treatise.
> II.—Improvement of the Understanding—Ethics—Letters.

HISTORY OF PHILOSOPHY.

BAX (E. B.). Handbook of the History of Philosophy. By E. BELFORT BAX. *2nd edition, revised.* Small post 8vo, 5*s.*

DRAPER (J. W.). A History of the Intellectual Development of Europe. By JOHN WILLIAM DRAPER, M.D., LL.D. With Index. 2 vols. Small post 8vo, 5*s.* each.

FALCKENBERG (R.). History of Modern Philosophy. By RICHARD FALCKENBERG, Professor of Philosophy in the University of Erlangen. Translated by Professor A. C. ARMSTRONG. Demy 8vo, 16*s.*

HEGEL'S Lectures on the Philosophy of History. Translated by J. SIBREE, M.A. Small post 8vo, 5*s.*

LAW AND POLITICAL ECONOMY.

KENT'S Commentary on International Law. Edited by J. T. ABDY, LL.D., Judge of County Courts and Law Professor at Gresham College, late Regius Professor of Laws in the University of Cambridge. *2nd edition, revised and brought down to a recent date.* Crown 8vo, 10*s.* 6*d.*

LAWRENCE (T. J.). Essays on some Disputed Questions in Modern International Law. By T. J. LAWRENCE, M.A., LL.M. *2nd edition, revised and enlarged.* Crown 8vo, 6*s.*

— Handbook of Public International Law. *2nd edition.* Fcap. 8vo, 3*s.*

MONTESQUIEU'S Spirit of Laws. A New Edition, revised and corrected, with D'Alembert's Analysis, Additional Notes, and a Memoir, by J. V. PRITCHARD, A.M. 2 vols. Small post 8vo, 3*s.* 6*d.* each.

PROTHERO (M.). Political Economy. By MICHAEL PROTHERO, M.A. Crown 8vo, 4*s.* 6*d.*

RICARDO on the Principles of Political Economy and Taxation. Edited by E. C. K. GONNER, M.A., Lecturer in University College, Liverpool. Small post 8vo, 5*s.*

SMITH (Adam). The Wealth of Nations. An Inquiry into the Nature and Causes of. Reprinted from the Sixth Edition, with an Introduction by ERNEST BELFORT BAX. 2 vols. Small post 8vo, 3*s.* 6*d.* each

HISTORY.

BOWES (A.). A Practical Synopsis of English History; or, A General Summary of Dates and Events. By ARTHUR BOWES. 10*th* *edition*. Revised and brought down to the present time. Demy 8vo, 1*s.*

COXE (W.). History of the House of Austria, 1218-1792. By ARCHDN. COXE, M.A., F.R.S. Together with a Continuation from the Accession of Francis I. to the Revolution of 1848. 4 vols. Small post 8vo. 3*s.* 6*d.* each.

DENTON (W.). England in the Fifteenth Century.. By the late REV. W. DENTON, M.A., Worcester College, Oxford. Demy 8vo, 12*s.*

DYER (Dr. T. H.). History of Modern Europe, from the Taking of Constantinople to the Establishment of the German Empire, A.D. 1453-1871. By DR. T. H. DYER. *A new edition*. In 5 vols. £2 12*s.* 6*d.*

GIBBON'S Decline and Fall of the Roman Empire. Complete and Unabridged, with Variorum Notes. Edited by an English Churchman. With 2 Maps. 7 vols. Small post 8vo, 3*s.* 6*d.* each.

GREGOROVIUS' History of the City of Rome in the Middle Ages. Translated by ANNIE HAMILTON. Vols. I., II., and III. Crown 8vo, 6*s.* each net.

GUIZOT'S History of the English Revolution of 1640. Translated by WILLIAM HAZLITT. Small post 8vo, 3*s.* 6*d.*
— History of Civilization, from the Fall of the Roman Empire to the French Revolution. Translated by WILLIAM HAZLITT. 3 vols. Small post 8vo, 3*s.* 6*d.* each.

HENDERSON (E. F.). Select Historical Documents of the Middle Ages. Including the most famous Charters relating to England, the Empire, the Church, etc., from the sixth to the fourteenth centuries. Translated and edited, with Introductions, by ERNEST F. HENDERSON, A.B., A.M., PH.D. Small post 8vo, 5*s.*
— A History of Germany in the Middle Ages. Post 8vo, 7*s.* 6*d.* net.

HOOPER (George). The Campaign of Sedan : The Downfall of the Second Empire, August-September, 1870. By GEORGE HOOPER. With General Map and Six Plans of Battle. Demy 8vo, 14*s.*
— Waterloo : The Downfall of the First Napoleon : a History of the Campaign of 1815. With Maps and Plans. Small post 8vo, 3*s.* 6*d.*

LAMARTINE'S History of the Girondists. Translated by H. T. RYDE. 3 vols. Small post 8vo, 3*s.* 6*d.* each.
— History of the Restoration of Monarchy in France (a Sequel to his History of the Girondists). 4 vols. Small post 8vo, 3*s.* 6*d.* each.
— History of the French Revolution of 1848. Small post 8vo, 3*s.* 6*d.*

LAPPENBERG'S History of England under the Anglo-Saxon Kings. Translated by the late B. THORPE, F.S.A. *New edition*, revised by E. C. OTTÉ. 2 vols. Small post 8vo, 3*s.* 6*d.* each.

MACHIAVELLI'S History of Florence, and of the Affairs of Italy from the Earliest Times to the Death of Lorenzo the Magnificent : together with the Prince, Savonarola, various Historical Tracts, and a Memoir of Machiavelli. Small post 8vo, 3*s.* 6*d.*

MARTINEAU (H.). History of England from 1800-15. By HARRIET MARTINEAU. Small post 8vo, 3*s.* 6*d.*

MARTINEAU (H.). History of the Thirty Years' Peace, 1815-46. 4 vols. Small post 8vo, 3s. 6d. each.

MAURICE (C. E.). The Revolutionary Movement of 1848-9 in Italy, Austria, Hungary, and Germany. With some Examination of the previous Thirty-three Years. By C. EDMUND MAURICE. With an engraved Frontispiece and other Illustrations. Demy 8vo, 16s.

MENZEL'S History of Germany, from the Earliest Period to 1842. 3 vols. Small post 8vo, 3s. 6d. each.

MICHELET'S History of the French Revolution from its earliest indications to the flight of the King in 1791. Small post 8vo, 3s. 6d.

MIGNET'S History of the French Revolution, from 1789 to 1814. Small post 8vo, 3s. 6d.

PARNELL (A.). The War of the Succession in Spain during the Reign of Queen Anne, 1702-1711. Based on Original Manuscripts and Contemporary Records. · By COL. THE HON. ARTHUR PARNELL, R.E. Demy 8vo, 14s. With Map, etc.

RANKE (L.). History of the Latin and Teutonic Nations, 1494-1514. Translated by P. A. ASHWORTH. Small post 8vo, 3s. 6d.

— History of the Popes, their Church and State, and especially of their conflicts with Protestantism in the 16th and 17th centuries. Translated by E. FOSTER. 3 vols. Small post 8vo, 3s. 6d. each.

— History of Servia and the Servian Revolution. Translated by MRS. KERR. Small post 8vo, 3s. 6d.

SIX OLD ENGLISH CHRONICLES: viz., Asser's Life of Alfred and the Chronicles of Ethelwerd, Gildas, Nennius, Geoffrey of Monmouth, and Richard of Cirencester. Edited, with Notes and Index, by J. A. GILES, D.C.L. Small post 8vo, 5s.

STRICKLAND (Agnes). The Lives of the Queens of England; from the Norman Conquest to the Reign of Queen Anne. By AGNES STRICKLAND. 6 vols. 5s. each.

— The Lives of the Queens of England. Abridged edition for the use of Schools and Families, Post 8vo, 6s. 6d.

THIERRY'S History of the Conquest of England by the Normans; its Causes, and its Consequences in England, Scotland, Ireland, and the Continent. Translated from the 7th Paris edition by WILLIAM HAZLITT. 2 vols. Small post 8vo, 3s. 6d. each.

WRIGHT (H. F.). The Intermediate History of England, with Notes, Supplements, Glossary, and a Mnemonic System. For Army and Civil Service Candidates. By H. F. WRIGHT, M.A., LL.M. Crown 8vo, 6s.

For other Works of value to Students of History, see Catalogue of
Bohn's Libraries, sent post-free on application.

DIVINITY, ETC.

ALFORD (DEAN). Greek Testament. With a Critically revised Text, a digest of Various Readings, Marginal References to verbal and idiomatic usage, Prolegomena, and a Critical and Exegetical Commentary. For the use of theological students and ministers. By the late HENRY ALFORD, D.D., Dean of Canterbury. 4 vols. 8vo. £5 2s. Sold separately.

— The New Testament for English Readers. Containing the Authorized Version, with additional Corrections of Readings and Renderings, Marginal References, and a Critical and Explanatory Commentary. In 2 vols. £2 14s. 6d. Also sold in 4 parts separately.

AUGUSTINE de Civitate Dei. Books XI. and XII. By the REV. HENRY D. GEE, B.D., F.S.A. I. Text only. 2s. II. Introduction and Translation. 3s.

— In Joannis Evangelium Tractates XXIV-XXVII. Edited by the REV. HENRY GEE, B.D., F.S.A. I. Text only, 1s. 6d. II. Translation by the late REV. CANON H. BROWN. 1s. 6d.

BARRETT (A. C.). Companion to the Greek Testament. By the late A. C. BARRETT, M.A., Caius College, Cambridge. 5th edition. Fcap. 8vo, 5s.

BARRY (BP.). Notes on the Catechism. For the use of Schools. By the RT. REV. BISHOP BARRY, D.D. 10th edition. Fcap. 2s.

BLEEK. Introduction to the Old Testament. By FRIEDRICH BLEEK. Edited by JOHANN BLEEK and ADOLF KAMPHAUSEN. Translated from the second edition of the German by G. H. VENABLES under the supervision of the REV. E. VENABLES, Residentiary Canon, of Lincoln. 2nd edition, with Corrections. With Index. 2 vols. small post 8vo, 5s. each.

BUTLER (BP.). Analogy of Religion. With Analytical Introduction and copious Index, by the late RT. REV. DR. STEERE. Fcap. 3s. 6d.

EUSEBIUS. Ecclesiastical History of Eusebius Pamphilus, Bishop of Cæsarea. Translated from the Greek by REV. C. F. CRUSE, M.A. With Notes, a Life of Eusebius, and Chronological Table. Sm. post 8vo, 5s.

GREGORY (DR.). Letters on the Evidences, Doctrines, and Duties of the Christian Religion. By DR. OLINTHUS GREGORY, F.R.A.S. Small post 8vo, 3s. 6d.

HUMPHRY (W. G.). Book of Common Prayer. An Historical and Explanatory Treatise on the. By W. G. HUMPHRY, B.D., late Fellow of Trinity College, Cambridge, Prebendary of St. Paul's, and Vicar of St. Martin's-in-the-Fields, Westminster. 6th edition. Fcap. 8vo, 2s. 6d. Cheap Edition, for Sunday School Teachers. 1s.

JOSEPHUS (FLAVIUS). The Works of. WHISTON's Translation. Revised by REV. A. R. SHILLETO, M.A. With Topographical and Geographical Notes by COLONEL SIR C. W. WILSON, K.C.B. 5 vols. 3s. 6d. each.

LUMBY (DR.). The History of the Creeds. I. Ante-Nicene. II. Nicene and Constantinopolitan. III. The Apostolic Creed. IV. The Quicunque, commonly called the Creed of St. Athanasius. By J. RAWSON LUMBY, D.D., Norrisian Professor of Divinity, Fellow of St. Catherine's College, and late Fellow of Magdalene College, Cambridge. 3rd edition, revised. Crown 8vo, 7s. 6d.

— Compendium of English Church History, from 1688-1830. With a Preface by J. RAWSON LUMBY, D.D. Crown 8vo, 6s.

MACMICHAEL (J. F.). The New Testament in Greek. With English Notes and Preface, Synopsis, and Chronological Tables. By the late REV. J. F. MACMICHAEL. Fcap. 8vo (730 pp.), 4s. 6d. Also the Four Gospels, and the Acts of the Apostles, separately. In paper wrappers, 6d. each.

MILLER (E.). Guide to the Textual Criticism of the New Testament. By REV. E MILLER, M.A., Oxon, Rector of Bucknell, Bicester. Crown 8vo, 4s.

NEANDER (DR. A.). History of the Christian Religion and Church. Translated by J. TORREY. 10 vols. small post 8vo, 3s. 6d. each.

— Life of Jesus Christ. Translated by J. McCLINTOCK and C. BLUMENTHAL. Small post 8vo, 3s. 6d.

— History of the Planting and Training of the Christian Church by the Apostles. Translated by J. E. RYLAND. 2 vols. 3s. 6d. each.

NEANDER (DR. A.). Lectures on the History of Christian Dogmas. Edited by DR. JACOBI. Translated by J. E. RYLAND. 2 vols. small post 8vo, 3*s*. 6*d*. each.

— Memorials of Christian Life in the Early and Middle Ages. Translated by J. E. RYLAND. Small post 8vo, 3*s*. 6*d*.

PEARSON (BP.). On the Creed. Carefully printed from an Early Edition. Edited by E. WALFORD, M.A. Post 8vo, 5*s*.

PEROWNE (BP.). The Book of Psalms. A New Translation, with Introductions and Notes, Critical and Explanatory. By the RIGHT REV. J. J. STEWART PEROWNE, D.D., Bishop of Worcester. 8vo. Vol. I. *8th edition, revised.* 18*s*. Vol. II. *7th edition, revised.* 16*s*.

— The Book of Psalms. Abridged Edition for Schools. Crown 8vo. *7th edition.* 10*s*. 6*d*.

SADLER (M. F.). The Church Teacher's Manual of Christian Instruction. Being the Church Catechism, Expanded and Explained in Question and Answer. For the use of the Clergyman, Parent, and Teacher. By the REV. M. F. SADLER, Prebendary of Wells, and Rector of Honiton. 43*rd thousand.* 2*s*. 6*d*.

.*. A Complete List of Prebendary Sadler's Works will be sent on application.

SCRIVENER (DR.). A Plain Introduction to the Criticism of the New Testament. With Forty-four Facsimiles from Ancient Manuscripts. For the use of Biblical Students. By the late F. H. SCRIVENER, M.A., D.C.L., LL.D., Prebendary of Exeter. *4th edition*, thoroughly revised, by the REV. E. MILLER, formerly Fellow and Tutor of New College, Oxford. 2 vols. demy 8vo, 32*s*.

— Novum Testamentum Græce, Textus Stephanici, 1550. Accedunt variae lectiones editionum Bezae, Elzeviri, Lachmanni, Tischendorfii, Tregellesii, curante F. H. A. SCRIVENER, A.M., D.C.L., LL.D. *Revised edition.* 4*s*. 6*d*.

— Novum Testamentum Græce [Editio Major] textus Stephanici, A.D. 1556. Cum variis lectionibus editionum Bezae, Elzeviri, Lachmanni, Tischendorfii, Tregellesii, Westcott-Hortii, versionis Anglicanæ emendatorum curante F. H. A. SCRIVENER, A.M., D.C.L., LL.D., accedunt parallela s. scripturæ loca. Small post 8vo. *2nd edition.* 7*s*. 6*d*.

An Edition on writing-paper, with margin for notes. 4to, half bound, 12*s*.

WHEATLEY. A Rational Illustration of the Book of Common Prayer. Being the Substance of everything Liturgical in Bishop Sparrow, Mr. L'Estrange, Dr. Comber, Dr. Nicholls, and all former Ritualist Commentators upon the same subject. Small post 8vo, 3*s*. 6*d*.

WHITAKER (C.). Rufinus and His Times. With the Text of his Commentary on the Apostles' Creed and a Translation. To which is added a Condensed History of the Creeds and Councils. By the REV. CHARLES WHITAKER, B.A., Vicar of Natland, Kendal. Demy 8vo, 5*s*.

Or in separate Parts.—1. Latin Text, with Various Readings, 2*s*. 6*d*. 2. Summary of the History of the Creeds, 1*s*. 6*d*. 3. Charts of the Heresies of the Times preceding Rufinus, and the First Four General Councils, 6*d*. each.

— St. Augustine: De Fide et Symbolo—Sermo ad Catechumenos. St. Leo ad Flavianum Epistola—Latin Text, with Literal Translation, Notes, and History of Creeds and Councils. 5*s*. Also separately, Literal Translation. 2*s*.

— Student's Help to the Prayer-Book. 3*s*.

SUMMARY OF SERIES.

BIBLIOTHECA CLASSICA.
PUBLIC SCHOOL SERIES.
CAMBRIDGE GREEK AND LATIN TEXTS.
CAMBRIDGE TEXTS WITH NOTES.
GRAMMAR SCHOOL CLASSICS.
PRIMARY CLASSICS.
BELL'S CLASSICAL TRANSLATIONS.
CAMBRIDGE MATHEMATICAL SERIES.
CAMBRIDGE SCHOOL AND COLLEGE TEXT BOOKS.
FOREIGN CLASSICS.
MODERN FRENCH AUTHORS.
MODERN GERMAN AUTHORS.
GOMBERT'S FRENCH DRAMA.
BELL'S MODERN TRANSLATIONS.
BELL'S ENGLISH CLASSICS.
HANDBOOKS OF ENGLISH LITERATURE.
TECHNOLOGICAL HANDBOOKS.
BELL'S AGRICULTURAL SERIES.
BELL'S READING BOOKS AND GEOGRAPHICAL READERS.

BIBLIOTHECA CLASSICA.

AESCHYLUS. By DR. PALEY. 8s.
CICERO. By G. LONG. Vols. I. and II. 8s. each.
DEMOSTHENES. By R. WHISTON. 2 Vols. 8s. each.
EURIPIDES. By DR. PALEY. Vols. II. and III. 8s. each.
HERODOTUS. By DR. BLAKESLEY. 2 Vols. 12s.
HESIOD. By DR. PALEY. 5s.
HOMER. By DR. PALEY. 2 Vols. 14s.
HORACE. By A. J. MACLEANE. 8s.
PLATO. Phaedrus. By DR. THOMPSON. 5s.
SOPHOCLES. Vol. I. By F. H. BLAYDES. 5s.
— Vol. II. By DR. PALEY. 6s.
VIRGIL. By CONINGTON AND NETTLESHIP. 3 Vols. 10s. 6d. each.

PUBLIC SCHOOL SERIES.

ARISTOPHANES. Peace. By DR. PALEY. 2s. 6d.
— Acharnians. By DR. PALEY. 2s. 6d.
— Frogs. By DR. PALEY. 2s. 6d.
CICERO. Letters to Atticus. Book I. By A. PRETOR. 4s. 6d.
DEMOSTHENES. De Falsa Legatione. By R. SHILLETO. 6s.
— Adv. Leptinem. By B. W. BEATSON. 3s. 6d.
LIVY. Books XXI. and XXII. By L. D. DOWDALL. 2s. each.
PLATO. Apology of Socrates and Crito. By DR. W. WAGNER. 3s. 6d. and 2s. 6d.
— Phaedo. By DR. W. WAGNER. 5s. 6d.
— Protagoras. By W. WAYTE. 4s. 6d.
— Gorgias. By DR. THOMPSON. 6s.
— Euthyphro. By G. H. WELLS. 3s.
— Euthydemus. By G. H. WELLS. 4s.
— Republic. By G. H. WELLS. 5s.
PLAUTUS. Aulularia. By DR. W. WAGNER. 4s. 6d.
— Trinummus. By DR. W. WAGNER. 4s. 6d.
— Menaechmei. By DR. W. WAGNER. 4s. 6d.
— Mostellaria. By E. A. SONNENSCHEIN. 5s.

PUBLIC SCHOOL SERIES—*continued.*

SOPHOCLES. Trachiniae. By A. PRETOR. 4s. 6d.
— Oedipus Tyrannus. By B. H. KENNEDY. 2s. 6d.
TERENCE. By DR. W. WAGNER. 7s. 6d.
THEOCRITUS. By DR. PALEY. 4s. 6d.
THUCYDIDES. Book VI. By T. W. DOUGAN. 2s.

CAMBRIDGE GREEK AND LATIN TEXTS.

AESCHYLUS. By DR. PALEY. 2s.
CAESAR. By G. LONG. 1s. 6d.
CICERO. De Senectute, de Amicitia, et Epistolae Selectae. By G. LONG. 1s. 6d.
— Orationes in Verrem. By G. LONG. 2s. 6d.
EURIPIDES. By DR. PALEY. 3 Vols. 2s. each.
HERODOTUS. By DR. BLAKESLEY. 2 Vols. 2s. 6d. each.
HOMER'S Iliad. By DR. PALEY. 1s. 6d.
HORACE. By A. J. MACLEANE. 1s. 6d.
JUVENAL AND PERSIUS. By A. J. MACLEANE. 1s. 6d.
LUCRETIUS. By H. A. J. MUNRO. 2s.
SALLUST. By G. LONG. 1s. 6d.
SOPHOCLES. By DR. PALEY. 2s. 6d.
TERENCE. By DR. W. WAGNER. 2s.
THUCYDIDES. By DR. DONALDSON. 2 Vols. 2s. each.
VIRGIL. By PROF. CONINGTON. 2s.
XENOPHON. By J. F. MACMICHAEL. 1s. 6d.
NOVUM TESTAMENTUM GRAECE. By DR. SCRIVENER. 4s. 6d.

CAMBRIDGE TEXTS WITH NOTES.

AESCHYLUS. By DR. PALEY. 6 Vols. 1s. 6d. each.
EURIPIDES. By DR. PALEY. 13 Vols. (Ion, 2s.) 1s. 6d. each.
HOMER'S Iliad. By DR. PALEY. 1s.
SOPHOCLES. By DR. PALEY. 5 Vols. 1s. 6d. each.
XENOPHON. Hellenica. By REV. L. D. DOWDALL. Books I. and II. 2s. each.
— Anabasis. By J. F. MACMICHAEL. 6 Vols. 1s. 6d. each.
CICERO. De Senectute, de Amicitia, et Epistolae Selectae. By G. LONG. 3 Vols. 1s. 6d. each.
OVID. Selections. By A. J. MACLEANE. 1s. 6d.
— Fasti. By DR. PALEY. 3 Vols. 2s. each.
TERENCE. By DR. W. WAGNER. 4 Vols. 1s. 6d. each.
VIRGIL. By PROF. CONINGTON. 12 Vols. 1s. 6d. each.

GRAMMAR SCHOOL CLASSICS.

CAESAR, De Bello Gallico. By G. LONG. 4s., or in 3 parts, 1s. 6d. each.
CATULLUS, TIBULLUS, and PROPERTIUS. By A. H. WRATISLAW, and F. N. SUTTON. 2s. 6d.
CORNELIUS NEPOS. By J. F. MACMICHAEL. 2s.
CICERO. De Senectute, De Amicitia, and Select Epistles. By G. LONG. 3s.
HOMER. Iliad. By DR. PALEY. Books I.-XII. 4s. 6d., or in 2 Parts, 2s. 6d. each.
HORACE. By A. J. MACLEANE. 3s. 6d., or in 2 Parts, 2s. each.
JUVENAL. By HERMAN PRIOR. 3s. 6d.
MARTIAL. By DR. PALEY and W. H. STONE. 4s. 6d.
OVID. Fasti. By DR. PALEY. 3s. 6d., or in 3 Parts, 1s. 6d. each.
SALLUST. Catilina and Jugurtha. By G. LONG and J. G. FRAZER. 3s. 6d., or in 2 Parts, 2s. each.
TACITUS. Germania and Agricola. By P. FROST. 2s. 6d.
VIRGIL. CONINGTON's edition abridged. 2 Vols. 4s. 6d. each, or in 9 Parts, 1s. 6d. each.
— Bucolics and Georgics. CONINGTON's edition abridged. 3s.
XENOPHON. By J. F. MACMICHAEL. 3s. 6d., or in 4 Parts, 1s. 6d. each.
— Cyropaedia. By G. M. GORHAM. 3s. 6d., or in 2 Parts, 1s. 6d. each.
— Memorabilia. By PERCIVAL FROST. 3s.

PRIMARY CLASSICS.

EASY SELECTIONS FROM CAESAR. By A. M. M. STEDMAN. 1s.
EASY SELECTIONS FROM LIVY. By A. M. M. STEDMAN. 1s. 6d.
EASY SELECTIONS FROM HERODOTUS. By A. G. LIDDELL. 1s. 6d.

BELL'S CLASSICAL TRANSLATIONS.

AESCHYLUS. By WALTER HEADLAM. 6 Vols. [*In the press.*
ARISTOPHANES. Acharnians. By W. H. COVINGTON. 1*s*.
CAESAR'S Gallic War. By W. A. MCDEVITTE. 2 Vols. 1*s*. each.
CICERO. Friendship and Old Age. By G. H. WELLS. 1*s*.
DEMOSTHENES. On the Crown. By C. RANN KENNEDY. 1*s*.
EURIPIDES. 14 Vols. By E. P. COLERIDGE. 1*s*. each.
HORACE. The Odes and Satires. By A. HAMILTON BRYCE, LL.D.
  [*In the press.*
LIVY. Books I.-IV. By J. H. FREESE. 1*s*. each.
— Book V. and VI. By E. S. WEYMOUTH. 1*s*. each.
— Book IX. By F. STORR. 1*s*.
LUCAN: The Pharsalia. Book I. By F. CONWAY. 1*s*.
OVID. Fasti. 3 Vols. By H. T. RILEY. 1*s*. each.
— Tristia. By H. T. RILEY. 1*s*.
SOPHOCLES. 7 Vols. By E. P. COLERIDGE. 1*s*. each
VIRGIL. 6 Vols. By A. HAMILTON BRYCE. 1*s*. each.
XENOPHON. Anabasis. 3 Vols. By J. S. WATSON. 1*s*. each.
— Hellenics. Books I. and II. By H. DALE. 1*s*.

CAMBRIDGE MATHEMATICAL SERIES.

ARITHMETIC. By C. PENDLEBURY. 4*s*. 6*d*., or in 2 Parts, 2*s*. 6*d*. each.
  Key to Part II. 7*s*. 6*d*. net.
EXAMPLES IN ARITHMETIC. By C. PENDLEBURY. 3*s*.. or in 2 Parts,
  1*s*. 6*d*. and 2*s*.
ARITHMETIC FOR INDIAN SCHOOLS. By PENDLEBURY and TAIT. 3*s*.
ELEMENTARY ALGEBRA. By J. T. HATHORNTHWAITE. 2*s*.
CHOICE AND CHANCE. By W. A. WHITWORTH. 6*s*.
EUCLID. By H. DEIGHTON. 4*s*. 6*d*., or Books I.-IV., 3*s*. : Books V.-XI., 2*s*. 6*d*. ;
  or Book I., 1*s*.; Books I. and II.. 1*s*. 6*d*.; Books I.-III., 2*s*. 6*d*.; Books III.
  and IV., 1*s*. 6*d*. KEY. 5*s*. net.
EXERCISES ON EUCLID, &c. By J. MCDOWELL. 6*s*.
ELEMENTARY MENSURATION. By B. T. MOORE.
ELEMENTARY TRIGONOMETRY. By C. PENDLEBURY. 4*s*. 6*d*.
ELEMENTARY TRIGONOMETRY. By DYER and WHITCOMBE. 4*s*. 6*d*.
PLANE TRIGONOMETRY. By T. G. VYVYAN. 3*s*. 6*d*.
ANALYTICAL GEOMETRY FOR BEGINNERS Part I. By T. G.
  VYVYAN. 2*s*. 6*d*.
ELEMENTARY GEOMETRY OF CONICS. . By DR. TAYLOR. 4*s*. 6*d*.
GEOMETRICAL CONIC SECTIONS. By DR. W. H. BESANT. 4*s*. 6*d*.
  Key. 5*s*. net.
GEOMETRICAL CONIC SECTIONS. By H. G. WILLIS. 5*s*.
SOLID GEOMETRY. By W. S. ALDIS. 6*s*.
GEOMETRICAL OPTICS. By W. S. ALDIS. 4*s*.
ROULETTES AND GLISSETTES. By DR. W. H. BESANT. 5*s*.
ELEMENTARY HYDROSTATICS. By DR. W. H. BESANT. 4*s*. 6*d*.
  Solutions. 5*s*. net
HYDROMECHANICS. Part I. Hydrostatics. By DR. W. H. BESANT. 5*s*.
DYNAMICS. By DR. W. H. BESANT. 10*s*. 6*d*.
RIGID DYNAMICS. By W. S. ALDIS. 4*s*.
ELEMENTARY DYNAMICS. By DR. W. GARNETT. 6*s*.
ELEMENTARY TREATISE ON HEAT. By DR. W. GARNETT. 4*s*. 6*d*.
ELEMENTS OF APPLIED MATHEMATICS. By C. M. JESSOP. 6*s*.
PROBLEMS IN ELEMENTARY MECHANICS. By W. WALTON. 6*s*.
EXAMPLES IN ELEMENTARY PHYSICS. By W. GALLATLY. 4*s*.
MATHEMATICAL EXAMPLES. By DYER and PROWDE SMITH. 6*s*.

CAMBRIDGE SCHOOL AND COLLEGE TEXT BOOKS.

ARITHMETIC. By C. ELSEE. 3*s*. 6*d*.
  By A. WRIGLEY. 3*s*. 6*d*.
EXAMPLES IN ARITHMETIC. By WATSON and GOUDIE. 2*s*. 6*d*.
ALGEBRA By C. ELSEE. 4*s*.
EXAMPLES IN ALGEBRA. By MACMICHAEL and PROWDE SMITH. 3*s*. 6*d*.
  and 4*s*. 6*d*.
PLANE ASTRONOMY. By P. T. MAIN. 4*s*.

CAMBRIDGE SCHOOL TEXTS—*continued.*

STATICS. By BISHOP GOODWIN. 3*s.*
NEWTON'S Principia. By EVANS and MAIN. 4*s.*
ANALYTICAL GEOMETRY. By T. G. VYVVAN. 4*s. 6d.*
COMPANION TO THE GREEK TESTAMENT. By A. C. BARRETT. 5*s.*
TREATISE ON THE BOOK OF COMMON PRAYER. By W. G. HUMPHRY. 2*s. 6d.*
TEXT BOOK OF MUSIC. By H. C. BANISTER. 5*s.*
CONCISE HISTORY OF MUSIC. By DR. H. G. BONAVIA HUNT. 3*s. 6d.*

FOREIGN CLASSICS.

FÉNELON'S Télémaque. By C. J. DELILLE. 2*s. 6d.*
LA FONTAINE'S Select Fables. By F. E. A. GASC. 1*s. 6d.*
LAMARTINE'S Le Tailleur de Pierres de Saint-Point. By J. BOÏELLE. 1*s. 6d.*
SAINTINE'S Picciola. By DR. DUBUC. 1*s. 6d.*
VOLTAIRE'S Charles XII. By L. DIREY. 1*s. 6d.*
GERMAN BALLADS. By C. L. BIELEFELD. 1*s. 6d.*
GOETHE'S Hermann und Dorothea. By E. BELL and E. WÖLFEL. 1*s. 6d.*
SCHILLER'S Wallenstein. By DR. BUCHHEIM. 5*s.*, or in 2 Parts, 2*s. 6d.* each.
— Maid of Orleans. By DR. W. WAGNER. 1*s. 6d.*
— Maria Stuart. By V. KASTNER. 1*s. 6d.*

MODERN FRENCH AUTHORS.

BALZAC'S Ursule Mirouët. By J. BOÏELLE. 3*s.*
CLARÉTIE'S Pierrille. By J. BOÏELLE. 2*s. 6d.*
DAUDET'S La Belle Nivernaise. By J. BOÏELLE. 2*s.*
GREVILLE'S Le Moulin Frappier. By J. BOÏELLE. 3*s.*
HUGO'S Bug Jargal. By J. BOÏELLE. 3*s.*

MODERN GERMAN AUTHORS.

HEY'S Fabeln für Kinder. By PROF. LANGE. 1*s. 6d.*
— — with Phonetic Transcription of Text, &c. 2*s.*
FREYTAG'S Soll und Haben. By W. H. CRUMP. 2*s. 6d.*
BENEDIX'S Doktor Wespe. By PROF. LANGE. 2*s. 6d.*
HOFFMANN'S Meister Martin. By PROF. LANGE. 1*s. 6d.*
HEYSE'S Hans Lange. By A. A. MACDONELL. 2*s.*
AUERBACH'S Auf Wache, and Roquette's Der Gefrorene Kuss. By A. A. MACDONELL. 2*s.*
MOSER'S Der Bibliothekar. By PROF. LANGE. 2*s.*
EBERS' Eine Frage. By F. STORR. 2*s.*
FREYTAG'S Die Journalisten. By PROF. LANGE. 2*s. 6d.*
GUTZKOW'S Zopf und Schwert. By PROF. LANGE. 2*s. 6d.*
GERMAN EPIC TALES. By DR. KARL NEUHAUS. 2*s. 6d.*
SCHEFFEL'S Ekkehard. By DR. H. HAGER. 3*s.*

The following Series are given in full in the body of the Catalogue.

GOMBERT'S French Drama. *See page* 31.
BELL'S Modern Translations. *See page* 34.
BELL'S English Classics. *See pp.* 24, 25.
HANDBOOKS OF ENGLISH LITERATURE. *See page* 26.
TECHNOLOGICAL HANDBOOKS. *See page* 37.
BELL'S Agricultural Series. *See page* 36.
BELL'S Reading Books and Geographical Readers. *See pp.* 25, 26.

CHISWICK PRESS:—C. WHITTINGHAM AND CO., TOOKS COURT, CHANCERY LANE.